# Creation and Restoration of Coastal Plant Communities

Editor

## Roy R. Lewis, III

Professor of Biology
Department of Biology
Hillsborough Community College
President
Mangrove Systems, Inc.
Tampa, Florida

CRC Press, Inc.
Boca Raton, Florida

Library of Congress Cataloging in Publication Data
Main entry under title:

Creation and restoration of coastal plant
    communities.

    Bibliography: p.
    Includes index.
    1. Coastal flora. 2. Tidemarsh flora.
3. Plant communities. 4. Revegetation.
5. Wildlife habitat improvement. 6. Coastal
zone management. I. Lewis, Roy R., 1944-
QK938.C6C73          639.9'9          82-1238
ISBN 0-8493-6573-2             AACR2

This book represents information obtained from authentic and highly regarded sources. Reprinted material is quoted with permission, and sources are indicated. A wide variety of references are listed. Every reasonable effort has been made to give reliable data and information, but the author and the publisher cannot assume responsibility for the validity of all materials or for the consequences of their use.

Direct all inquiries to CRC Press, Inc., 2000 Corporate Blvd., N.W., Boca Raton, Florida, 33431.

© 1982 by CRC Press, Inc.

International Standard Book Number 0-8493-6573-2

Library of Congress Card Number 82-1238
Printed in the United States

# PREFACE

Ecologists can learn a great deal from damaged ecosystems, and possibly even more from attempts to restore and manage them. Doing so, however, will require closer cooperation between ecologists, field people, and industry.

John Cairns, Jr.[1]

The creation and restoration of habitats or ecosystems has been the subject of a number of recent symposia.[2-14] The field, however, is a rapidly expanding one that has recently lead to the initiation of the journal *Restoration and Management Notes* from which John Cairns quote was taken.

In an attempt to encourage information transfer and cooperation between all interested parties, I was asked to bring together all available information on creation and restoration of coastal plant communities in the form of synthesis papers and other appropriate information. (See Appendixes I and II). The following series of papers and their associated literature reviews were intended to provide the reader with a single source volume from which to draw all literature sources available on this subject. Since much of the literature discussed here is difficult to secure, we invite requests for assistance in locating information once the normal channels are exhausted.

I had asked Dr. Derek S. Ranwell to provide a chapter on European marsh systems, but he had to decline due to other commitments. He has advised me that he is preparing his own updated summaries of information for that region and readers are advised to write to him directly (see address in Appendix I) in addition to consulting his publications.[15]

As this book goes to press, additional projects are being initiated and others finished. An example of the latter is the work of Dr. Ernest Seneca in France where Coastal marshes have been restored following the *Amoco Cadiz* oil spill. In order to keep up with this field, readers are urged to consult the information in Appendixes I and II and use it.

I welcome communications from anyone reading this book concerning sources of information we may not have included and results of projects not reported here. We will make sure they are included in any further additions of this book.

R. R. Lewis, III
Tampa, Florida
December, 1981

# REFERENCES

1. Cairns, J., Jr., Restoration and management: an ecologist's perspective, *Restor. Manage. Notes,* 1(1), 6, 1981.
2. Cairns, J., Jr., Dickson, K. L., and Herricks, E. E., Eds., Recovery and restoration of damaged ecosystems, *Proc. Int. Symp. Recovery Damaged Ecosystems,* Virginia Polytechnic Institute and State University, Blacksburg, March 23—25, 1975, University Press of Virginia, Charlottesville, 1977, 531.
3. Holdgate, M. W. and Woodman, M. J., Eds., The breakdown and restoration of ecosystems, *Proc. Conf. Rehab. Severely Damaged Land Freshwater Ecosystems Temperate Zones,* Reykjavik, Iceland, July 4—10, 1976, Plenum Press, New York, 1978, 496.
4. Cairns, J., Jr., Ed., *The Recovery Process in Damaged Ecosystems,* Ann Arbor Science, Ann Arbor, Mich., 1980, 167.
5. Lewis, R. R. and Thomas, J., Eds., *Proc. 1st Annu. Conf. Restor. Coastal Vegetation Fla.,* May 4, 1974, Hillsborough Community College, Tampa, Fla., Florida Audubon Society, Maitland, 1975, 55.
6. Lewis, R. R., Ed., *Proc. 2nd Annu. Conf. Restor. Coastal Vegetation Fla.,* May 17, 1975, Hillsborough Community College, Tampa, Fla., 1976, 203.
7. Lewis, R. R. and Cole, D. P., Eds., *Proc. 3rd Annu. Conf. Restor. Coastal Vegetation Fla.,* May 15, 1976, Hillsborough Community College, Tampa, Fla., 1977, 147.
8. Lewis, R. R. and Cole, D. P., *Proc. 4th Annu. Conf. Restor. Coastal Vegetation Fla.,* May 14, 1977, Hillsborough Community College, Tampa, Fla., 1978, 165.
9. Lewis, R. R. and Cole, D. P., Eds., *Proc. 5th Annu. Conf. Restor. Coastal Vegetation Fla.,* May 13, 1978, Hillsborough Community College, Tampa, Fla., 1979, 255.
10. Cole, D. P., Ed., *Proc. 6th Annu. Conf. Wetlands Restor. Creation,* May 19, 1979, Hillsborough Community College, Tampa, Fla., 1980, 356.
11. Cole, D. P., Ed., *Proc. 7th Annu. Conf. Restor. Creation Wetlands,* May 16—17, 1970, Hillsborough Community College, Tampa, Fla., 1981, 294.
12. Stovall, R., Ed., Proc. 8th Annu. Conf. Restor. Creation Wetlands, May 8—9, Hillsborough Community College, Tampa, Fla., 1981, 200.
13. Swanson, G. A., Tech. Coord., The mitigation symposium: a national workshop on mitigating losses of fish and wildlife habitats, July 16—20, 1979, Gen. Tech. Rep. RM-65, Rocky Mountain Forest and Range Exp. Stn., Ft. Collins, Colo., 1979, 684.
14. Thorhaug, A., Symposium on Restoration of Major Plant Communities in the United States, June, 1976, Reprints published from *Env. Conserv.,* various issues, Elsevier, Lausanne, Switzerland.
15. Ranwell, D. S., *Ecology of Salt Marshes and Sand Dunes,* Chapman and Hill, London, 1972, 258.

# THE EDITOR

Roy R. "Robin" Lewis, III, M.A., is President of Mangrove Systems, Inc., an environmental consulting company located in Tampa, Fla. He is also a Senior Adjunct Scientist with the Mote Marine Laboratory, Sarasota, Fla.

Mr. Lewis was graduated from the University of Florida, Gainesville, with a B.Sc. degree in Biology in 1966. He received a M.A. degree from the University of South Florida in Zoology in 1968 and did postgraduate work in Marine Science at the University of South Florida main campus and at the Marine Science Institute in St. Petersburg, Fla. between 1969 and 1973. Mr. Lewis was an Instructor (1969 to 1970), Assistant Professor (1970 to 1972), Associate Professor (1972 to 1974), and finally Professor of Biology (1974 to 1982) at Hillsborough Community College, Tampa, Fla. He also served as Chairman oi the Department of Biology at this same institution between 1974 and 1977.

Mr. Lewis has published more than 40 papers on the ecology, restoration and creation of tidal marshes, mangrove forests, and seagrass meadows. Mr. Lewis is a member of more than a dozen professional societies including the International Association of Aquatic Vascular Plant Biologists, International Society of Tropical Forestry, Association for Tropical Biology, Ecological Society of America, Coast and Wetlands Society (Australia), and Society of Wetland Scientists where he serves as a member of the editorial board of the journal of the society, *Wetlands*.

Mr. Lewis has been the convenor and editor or editorial advisor of proceedings of the annual "Conferences on Creation and Restoration of Wetlands" now in their 9th year.

# CONTRIBUTORS

**Chung-Hsin Chung**
Director
Institute of Spartina and Tidal Land
  Development
Professor of Plant Ecology
Graduate School
University of Nanking
Nanking, China

**Paul L. Knutson**
Coastal Ecologist
U.S. Army Corps of Engineers
Coastal Engineering Research Center
Fort Belvoir, Virginia

**William L. Kruczynski**
Life Scientist
Ecological Review Branch
U.S. Environmental Protection Agency
Atlanta, Georgia

**Roy R. Lewis, III**
Professor of Biology
Hillsborough Community College
President
Mangrove Systems, Inc.
Tampa, Florida

**Ronald C. Phillips**
Professor of Biology
School of Natural and Mathematical
  Sciences
Seattle Pacific University
Seattle, Washington

**James W. Webb, Jr.**
Assistant Professor
Marine Biology Department
Texas A&M University at Galveston
Galveston, Texas

**W. W. Woodhouse, Jr.**
Professor Emeritus
North Carolina State University
President
Dune and Marsh, Inc.
Raleigh, North Carolina

# TABLE OF CONTENTS

Chapter 1

# COASTAL SAND DUNES OF THE U.S.

### W. W. Woodhouse, Jr.

## TABLE OF CONTENTS

# I. BEACH AND DUNE COMMUNITIES

Vegetated dunes occur on sandy seacoasts throughout the world in regions receiving enough rainfall to support plant growth and sufficient wind action to move sand. This includes a wide range of climates, from humid to semiarid, from cold to tropical. These dunes develop through the trapping of wind-blown sand by vegetation having special adaptations for sand accumulation. Coastal dunes vary widely in size, shape, and location depending upon sand supply and climate. In the absence of stabilizing vegetation, blowing sand often drifts into large ''live'' dunes that move back and forth with the wind (Figure 1).

Typically, the dune complex divides roughly into three zones. In order of increasing distance from the sea and in order of increasing age, these are (1) the pioneer zone, (2) the intermediate or scrub zone, and (3) the back dune or forest zone.

The location, extent, and nature of these zones vary widely depending on a number of factors such as coastal topography, climate, nature and rate of erosion and deposition, and sea level changes. The pioneer zone is sometimes lacking due to recent severe storms or to persistent long-term beach recession and in extreme cases both the pioneer and intermediate zones may be lost leaving the forest zone next to the beach. Also, on very high coasts (where salt spray is confined near the surf zone) forests may develop close to the sea.

Dune sands are basically unstable materials, readily moved and shaped by wind and water action. Consequently, disturbed dunes revert rapidly to pioneer zone conditions, regardless of their stage of development at the time of disturbance. Loss of vegetative cover immediately interrupts soil development processes and accumulated organic matter is soon dissipated. Many large live dunes and dune fields in coastal regions owe their origin to the removal of protective cover from stable dunes of long standing through fire, lumbering, overgrazing, or foot and vehicular traffic.[1-3] Consequently, dune restoration must almost always begin with the reinitiation of the whole plant succession sequence for that region through the establishment of pioneer zone species. For this reason, attention will be focused primarily on pioneer plants (Figure 2).

Direct planting of intermediate zone species under pioneer zone conditions has seldom been successful and little is known about their propagation. They usually invade such areas rapidly as soon as conditions permit. Extensive forest plantings have been carried out on large dune fields previously stabilized by grasses, forbs, and shrubs.[2,4] Woodhouse[26] recently reviewed the available information on dune vegetation establishment and restoration. This chapter is largely an adaptation of this report.

## A. Vegetation Zones
### 1. Pioneer Zone

This is the area of recent or continuing sand movement that usually occurs on the upper beach and foredunes. It is wide on prograding beaches, less so on stable sites, and narrow to nonexistent on receding coasts. Vegetation in a typical pioneer zone is limited to a few species of grasses, sedges, and forbs that can withstand salt spray, sandblast, sand burial, flooding, drought, as well as wide temperature fluctuations and low nutrient supply.

Although pioneer species for any given region tend to be few in number, individual species often play distinctly different roles. Some might be termed ''dune initiators'' and others ''dune builders'' while the majority act largely as stabilizers. Along the Atlantic and Gulf Coasts, sea purslane (*Sesuvium portulacastrum*) is a good example of the first group. It is a low-growing, spreading type occurring on dunes but capable of invading the exposed upper beach where it may form low, embryonic dunelets. These offer a toe hold to the dune builders by trapping their seeds and protecting them through seedling development. Dogs tail grass is another example of this type of plant common to the Dutch coast. Sea rocket (*Cakile maritima*) is a very widely adapted annual, occurring on dunes and upper beach through much of the world. The stiff stems of this species often trap debris, which may include seeds of dune builders.

The primary dune builders, usually perennial grasses, are very few in number, but generally adapted over very wide geographic ranges. European beachgrass or marram

FIGURE 1.    Live dune encroaching on forest, Kill Devil Hills, N.C.

FIGURE 2.    Dune developing around a dune grass planting in pioneer zone.

grass (*Ammophilia arenaria*) is the primary dune builder on the coasts of North Western Europe, the Pacific Northwest of North America, and elsewhere. Another *Ammophilia* species (*breviligulata*), American beachgrass, dominates the dune building process along the North and Mid-Atlantic coasts of North America and is very effective when planted along the Pacific Coast and elsewhere. Sea oats (*Uniola paniculata*) is the principal dune builder along most of the Atlantic and Gulf coasts of North

America from the Carolinas to Mexico with bitter panicum (*Panicum amarum*) becoming increasingly prominent in the southern half of this region. Sand spiniflex grass plays this role on the Queensland coast of Australia.

There are numerous species that invade the foredunes and assist in the stabilizing process. These include some of the aforementioned dune initiators such as sea purslane and sea rocket and a much larger variety of secondary invaders. Some of the more common are *Abrona, Ambrosia, Artemesia, Croton, Carex, Carpobrotus, Fimbristylis, Euphorbia, Erigeron, Fimbristilis, Festuca, Ipomoea, Lathyrus, Lupinus, Haplopappus, Hydrocotyle, Iva, Schizachyrium, Soldago, Spartina,* and *Sporobolus.* Some of these, *Abrona, Carpobrotus,* and *Hydrocotyle,* for example, are dune builders under some circumstances but definitely less effective than the perennial grasses. Some annuals such as *Strophostyles* vary widely in abundance and distribution from year to year, being dependent upon appropriate timing of sand movement for seed burial.

### 2. Scrub or Intermediate Zone

This is a highly variable, ill-defined area lying immediately behind the more active pioneer zone. It consists of secondary dune ridges and swales, flats, deflation plains, and occasionally includes the back slopes of large foredunes. Plants in this zone may include (in addition to the pioneer species), forbs, shrubs, and in some cases, stunted trees. The area receives little fresh sand and nutrient levels are often low resulting in a scrubby, starved appearance of the vegetation. Growth and vigor, particularly of the remaining pioneer species, is substantially lower than in the active zone, but the intermediate zone plants are adapted to these conditions and are valuable as stabilizers. Some intermediate species are planted for ornamental purposes and as wind breaks. In nature, these plants tend to invade and gradually replace the pioneers as soon as an area becomes sufficiently stable and oceanic influence is sufficiently reduced. This is why these plants are not usually planted in the pioneer zone.

This zone is normally considered a progression in the ecological succession toward the stable climax forest, away from the highly changeable, unstable state of the pioneer zone. Sand movement decreases or ceases completely. The area is populated by a wider variety of plants and the processes of soil development start. This continues as long as protection from the sea is provided. However, from the standpoint of stabilization, the plant cover on the typical intermediate zone affords less protection against wind and water erosion than does the cover on a well-vegetated foredune in the pioneer zone. Many intermediate zone plants cannot withstand sand burial and the thin, open ground cover is less resistant to wind and water erosion. Where continued protection is required and especially where overtopping by storm tides is a threat, it may be desirable to delay the ecological succession and retain a protective cover of vigorous pioneer zone plants. Although the reasons for the decline in vigor of pioneer species as sand accumulation decreases are not well-understood, nutrient supply associated with the fresh sand can be a major factor. Some sand-starved, declining stands of both American beachgrass (*Ammophila breviligulita*) and sea oats (*Uniola paniculata*) have been revived through the application of fertilizers.[5] Fertilization has been reported as useful in maintaining plant vigor in diverse regions (Atlantic coast, Oregon, Texas, and Europe).

### 3. Forest Zone

Forests form on dunes only after a substantial period of soil development and only on sites with considerable protection from salt spray and flooding. The vegetation varies from the dense thickets of trees, shrubs, and vines of the maritime forests of the South Atlantic and Gulf coasts to the coniferous, hardwood, and mixed open tree forests of the Great Lakes dunes.

After dune stabilization with pioneer and intermediate zone species, trees have been planted successfully to convert large mobile dunes and dune fields to forest.[4,6] However, trees are not planted in the barrier dune environment and are not desirable where the dune protects the backshore. Trees will shade out the pioneer and intermediate species and leave the dune unprotected and unable to regenerate cover quickly following a severe disease, insect outbreak, or fire. Fire was probably a major factor in the initiation of large mobile dunes in the Pacific Northwest.

The major pioneer species extend over much wider ranges, geographically and climatically, than do the plants of the intermediate and forest zones. However, the number of species is usually much greater in the more stable areas.

### B. Productivity and Habitat Values

The dune environment is not conducive to high productivity. The substrate is usually very low in nutrients and quite drouthy. Plants are frequently subjected to stress from such factors as temperature extremes, salinity, sand blast, sand burial, flooding, etc. Except for the forest zone, where significant soil formation has occurred and there is more distance and protection from the rigors of the sea, general productivity is low.

Dunes provide valuable habitat for numerous species. A variety of birds and rodents are permanent residents. The dune complex has contained the primary nesting sites for several ground-nesting birds, particularly gulls and terns, having exacting cover type and isolation requirements. This community forms an essential link in the migratory pathways of a large variety of birds — from warblers to peragrine falcons, provides resting sites for many, and food for others. Above all, perhaps, for many it has afforded isolation.

Coastal dunes, particularly those of the barrier type, have a value well beyond that of habitat, serving as coastal protection and preservation in several ways. Continuous barrier dunes serve as flexible barriers to storm surges and waves and are of particular value in affording protection to low-lying backshore areas and in helping to preserve the integrity of low barrier islands (Figure 3). For example, adequately stabilized and carefully protected foredunes play a significant role in the sea defenses of Holland.[7] Where there is an adequate natural sand supply, dunes provide protection more effectively and at a lower cost than a seawall. Well-developed barrier dunes perform another important function by providing stockpiles of sand to nourish the beach during storm attack. Storm waves erode sand from the berm and foredune, much of which is deposited immediately offshore in a bar formation, which allows the beach profile to adjust to the energy of the storm. In the absence of the dune sand reservoir, sand for storm profile adjustments must come from either the shore behind or the beach. In a stable beach-dune system, sand removed from berm and dune by storms is returned to the beach and berm by calm weather between storms. Thus, it is available to rebuild the dune, a necessary process in maintaining a dynamic equilibrium within the beach-berm-dune system.

The use of dunes must be carefully controlled, and their management should always include provisions for repair and restoration, as required.

### C. Habitat Loss or Modification

Sandy seacoasts are basically unstable and their natural cycles of erosion and deposition result in both continual dune building and periodic destruction. Further, the general rise of the level of the sea in relation to the land that has been in progress, particularly along the Atlantic and Gulf Coasts, for centuries has destroyed dunes and caused the landward migration of others.

Data on the aereal extent of coastal dune communities, past or present, are scarce and unreliable. The extensive Provincetown dunes on Cape Cod are known to have

FIGURE 3.    Large stable barrier dune, Padre Island, Tex.

been drastically altered as early as 1714.[1] Large areas of stable dunes along the Oregon coast were destabilized early in the settlement of that region through grazing, fire, and cultivation. The resulting migrating dunes were the subject of an extensive planting effort in the 1930s.[4] Large ''live'' dunes are still present on that coast. Blowing sand from destabilized dunes was a problem in the vicinity of San Francisco early in this century. The deforestation of Lake Michigan dunes created acute sand migration problems, threatening farms and towns with inundation.[6]

Although documentation is fragmentary, there is little doubt that most of the sandy portions of the Atlantic, Gulf, and Pacific coasts were bordered by dunes in various sizes, shapes, and vegetative stages prior to colonization. Extensive destruction of dune communities began very early, primarily through deforestation, as on Cape Cod,[1] or grazing. The forest zone of the dune complex often grew valuable timber and this was usually located close to the waterways. The extensive dune grasslands could support herds of cattle, sheep, horses, goats, and even pigs. The barrier islands on which many of the dune grasslands occurred offered effective natural enclosures for livestock where they could be left to fend for themselves. The natural increase in animal numbers usually led to overgrazing. Dune vegetation is particularly vulnerable to damage from grazing and trampling. Paths cut by sharp hooves soon start blowouts and begin the unraveling process which produces mobile dunes. Grazing was still a problem on Padre Island as late as 1970.[8] Livestock were not removed from Portsmouth Island and Core Banks on the North Carolina coast until the 1950s. It appears that most coastal dune areas along the Atlantic and Gulf were subjected to grazing at one time or another, many to the point of severe alteration of the plant community. The same is probably true of many on the Pacific Coast.

The presence and activities of man in modern times has accelerated loss of dune communities. The effect of reacreational and commercial development, off-the-road vehicles, and the general population pressures associated with the attraction coasts have for people has brought on widespread alteration and destruction. The building of streets and structures near the sea disrupts normal beach and dune processes. The natural cycle of beach nourishment by dunes during storms followed by dune replenishment during mild weather can no longer operate. Also, such man-made alterations often leave no space for the normal movement of sandy seashores and prevent the landward development and movement of new dunes along receding shorelines.

Heavy usage of dune areas has had substantial effects upon the fragile dune plant communities. In some cases this has been followed by the invasion of exotic species, such as Hottentot sea fig or ice plant along the California coast, resulting in a community totally different from the original.

## D. Loss Impacts

The impact of habitat loss due to the disappearance of dune communities is difficult to document. For some species, the impact has to be substantial. One of the more obvious impacts has been upon nesting habits of certain bird species. Several of those that formerly nested in and around the dune community, largely gulls and terns, have abandoned these areas in favor of dredge spoil islands. The reasons for these shifts may be more a matter of the need for isolation and protection from predators than actual changes in the vegetation on the dunes. However, in some cases, it is also a case of the total loss of dune communities. Certainly, the loss of the protective and beach nourishment functions of barrier dunes has had extensive impacts on many shores.

Other impacts are less clear-cut being more of a qualitative nature, but nevertheless significant in their overall effects.

## II. PLANT TYPES AND SOURCES

### A. Coastal Regions

The coasts of the continental U.S. should be divided into regions for planting purposes. These regions deviate from those drawn along climatic boundaries. Climatic effects become somewhat blurred and distorted in the narrow band near the water where coastal dunes occur. The localized maritime influence on fog, humidity, frost, temperature fluctuations, etc. enables some pioneer species to range across a variety of climatic zones. Divisions along species adaptation lines are more useful in dealing with those plants.

**North Atlantic region** — The North Atlantic region extends from the Canadian border to the Virginia Capes with a shoreline of about 1660 km. For planting purposes, the mid-Atlantic region is combined with the North Atlantic region. American beachgrass is the dominant pioneer plant for this region.

**South Atlantic region** — This region covers a rather wide range, climatically, from the Virginia Capes to Key West, a shoreline distance of about 1900 km. The Atlantic coasts of central and southern Florida are arbitrarily included in this region, because their separation is not useful from the narrow view of pioneer zone planting and protection. Many subtropical and tropical plants are used in the southern half of the Atlantic coast of Florida near the beach for ornamental purposes but the dune-building species do not differ significantly from the remainder of the region. The northern boundary of this region coincides with the northern limit of sea oats. Sea oats is the dominant foredune plant of this region.

**Gulf of Mexico region** — This region includes the gulf coast of Florida and extends around the gulf to the Mexican border, a distance of about 2600 km, but about 500 to 600 km is marshy with no beach or dune development. Climate varies from humid to semiarid but pioneer species planted, primarily sea oats and bitter panicum (*Panicum amarum*), are the same throughout the region.

**North Pacific region** — This region extends from the Canadian border to Monterey, Calif., a distance of about 1450 km. The southern limit is based on the transition zone between the grass-dominated communities to the north and the forb-dominated communities to the south.[9] European beachgrass (*Ammophilia arenaria*) and American dunegrass (*Elymus mollis*) dominate this region.

**South Pacific region** — This region extends southward from Monterey, to the Mexican border for about 650 km of coast that has a pronounced decrease in rainfall from north to south. Dominant plants are forbs such as sea fig (*Carpobrotus* spp.), sagewort (*Artemisia pcynocephala*), beach bur (*Ambrosia* spp.), and sand verbena (*Abrona* spp.).

Great Lakes region — This includes all of the shores of the Great Lakes within the U.S. However, dune development is confined largely to the Michigan and Indiana shores of Lake Michigan. American beachgrass and Prairie sandreed (*Calamovilla longifolia*) are dominant.

## B. Major Pioneer Plants

Bare dunes and dune fields along the coasts of this country are usually stabilized by planting with a small group of pioneer plants, perennial dune grasses. The major grasses are European beachgrass on the North and South Pacific coasts; American dunegrass on the North Pacific coast; American beachgrass on the North Atlatic and Great Lakes Coasts; sea oats, salt-meadow cordgrass (*Spartina patens*), and bitter panicum (*Panicum amarum*) on the South Atlantic and Gulf coasts.

These plants are used because they multiply dependably and economically, can be readily harvested, transported, stored, and planted. The plants thrive in blowing sand, trap sand well, and are relatively free of serious pests. These plants live from one year to the next, providing year-round protection to the sand surface.

There are other plants that invade the pioneer zone and contribute substantially to the process in each geographical region. None are widely planted for initial stabilization because they fail to meet one or more of the above criteria. There are many plants that, given extra care, will grow in the pioneer zone for ornamental or other specialized purposes. Some of these may find wider use as more is known about their requirements. A few species, such as the omnipresent sea rocket (*Cakile* sp.), may precede the dune grasses into the pioneer zone, but are capable of building only embryonic dunes. These species give other plants a chance to build more substantial dunes by temporarily trapping sand, seeds, and debris.

### 1. American Beachgrass

This is a cool season dune grass native to the North and mid-Atlantic and Great Lakes coasts, and is probably the most widely used species for initial stilling of blowing sand. It is almost the only species planted for this purpose along the Atlantic coast south to the Carolinas and around the Great Lakes. Limited use has been made of this grass in the Pacific Northwest. American beachgrass is a vigorous, erect grass which grows in dense clumps and is capable of rapid lateral spread by rhizomes. It can usually be recognized in the fruiting stage by the dense, cylindrical spikes or seed heads (Figure 4 and 5). There are two varieties of American beachgrass: Cape, a vigorous coarse-stemmed type adapted to the North Atlantic coast, and Hatteras, a fine-leaf type selected for early vigor on the coasts of the Carolinas. Characteristics of American beachgrass are

1. Easy to multiply vegetatively (50-fold increase per year is possible), and readily available commercially.
2. Easy to harvest, store, and transplant manually or by machine.
3. Long transplanting season with good survival rate (normally 90%).
4. Grows rapidly and becomes an effective sand trapper by middle of the first growing season.
5. A cool weather grower that starts growth in early spring and, where conditions are favorable, continues well into the fall.
6. Spreads outward 3 to 4 m/year and produces the most gentle seaward dune slopes of all the species.[10]

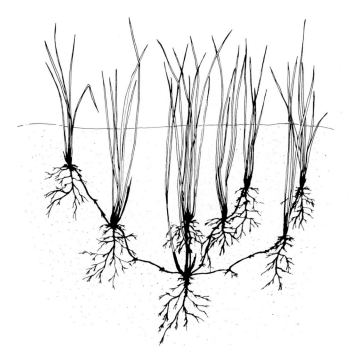

FIGURE 4. Growth habit and spread of American beachgrass under sand accumulation.

FIGURE 5. Morphological features of American beachgrass.

FIGURE 6.    Morphological features of European beachgrass.

7. Vigorous rhizome system makes it very effective at filling open stands and excellent for maintenance and repair of plantings.
8. Responds vigorously to fresh sand (will grow through 1.2 m of accumulating sand in one growing season; Figure 4).
9. Fades rapidly when sand and nutrient supply is cut off, and tends to be short-lived in the intermediate zone unless fertilized.
10. Affected by heat and drought in the southern part of its range where it is usually replaced within a few years by better adapted species such as sea oats and bitter panicum.
11. Very susceptible to a soft scale, *Eriococcus carolinae,*[11] throughout most of its Atlantic coast range and to Marasmius blight, a fungus pest,[12] along the South Atlantic coast.

## 2. European Beachgrass

A species very similar to American beachgrass (Figure 6), that was introduced in the late 1800s to the Pacific coast where it has become widely distributed by planting and natural spread. This is the marram grass of the British Isles and northern Europe, and probably the dune grass most extensively planted in the past. It is the principal grass used along the Pacific coast for the initial stabilization of large areas of blowing sand and the building of foredunes. Characteristics of European beachgrass are

1. Exceptionally easy to multiply vegetatively (a 100-fold increase per year is not unusual under nursery conditions), and available commercially.
2. Easy to harvest, store, and transplant, but will not tolerate as high a temperature as American beachgrass.
3. Long transplanting season, with excellent survival under proper conditions, and suitable for machine or manual transplanting.

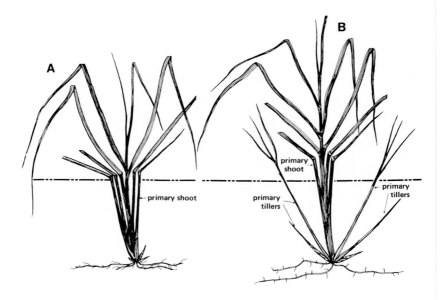

FIGURE 7.   Growth habit of sea oats. (A) Soon after planting, (B) after several months.

4.   Grows rapidly and responds vigorously to a plentiful sand supply (grows through as much as 60 cm of sand per year).
5.   Rhizomes shorter than those of American beachgrass apparently cause steeper seaward dune slopes (no documentation available) and are less effective in filling open stands for maintenance and repair, therefore, good initial stands are critical.
6.   Definitely a cool weather grass, very sensitive to high temperatures at transplanting; poor survival can be expected whenever air temperature exceeds 16°C during or immediately following transplanting; this characteristic may account for the lack of use of this plant on the Atlantic coast.
7.   Fertilizer application is essential for success on large mobile dunes.
8.   Vigor declines rapidly with stilling of sand, making it short-lived in the intermediate zone and requiring reinforcement by planting of intermediate species, particularly on large dunes.

*3. Sea Oats*

This is a native warm season dune grass occurring from about the Virginia Capes southward into Mexico and on some islands in the Carribean Sea. It is similar in appearance to American beachgrass, but with generally larger stems, more decumbent growth habit, and more open stands (Figures 7 and 8). The grass tends to dominate active foredunes throughout much of its range. Its striking appearance, particularly when in flower or fruit, has made legal protection necessary in some states to avoid excessive harvest for ornamental purposes. Characteristics of sea oats are

1.   Difficult and slow to multiply vegetatively, less than half as fast as American beachgrass; subject to pest problems when grown away from the beach; commercial availability is very limited.
2.   Spread into dune and beach areas is primarily by seeds; plants can be grown from seeds but are subject to pest problems in the field; grass is an erratic seed producer and seed heads are heavily preyed upon by insects, birds, rodents, and people.

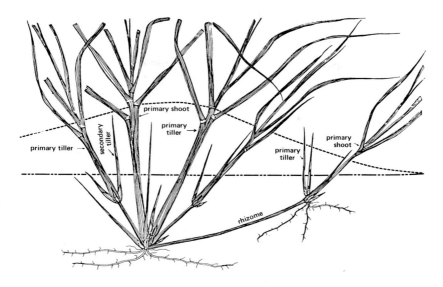

FIGURE 8. Sea oats at end of first summer.

3. Early growth of transplants and seedlings is very slow and survival is erratic, consequently, plantings of pure stands usually are only marginally effective as sand trappers during the year of establishment; therefore, it should rarely be planted alone. Mixed with bitter panicum (southern range) and American beachgrass or bitter panicum or both (northern range), sea oats will tend to increase with time.

4. Trapping capacity develops rapidly after the first year and even very spotty stands become effective with time.

5. It is more tolerant of reduced sand and nutrient supply than American beachgrass; therefore, it persists longer and provides more cover in the intermediate zone.

6. This grass is relatively free of pests in the dune environment.

7. The stiff upright growth habit and short rhizome system produce rapid sand accumulation near margins of vegetation. The slow lateral spread produces steep seaward dune slopes.

8. A summer grower that grows only during warm weather, which makes the growing season short in the Carolinas but almost year round on the south Texas coast.

9. Tolerance to heat and drought is high, and rarely exhibits drought symptoms.

### 4. Bitter Panicum

This is a strongly rhizomatous perennial warm season grass which spreads by branching at the nodes (Figures 9 and 10). It is generally less erect and has shorter leaves than sea oats and the beachgrasses. Bitter panicum has received serious attention only recently. It is apparently native to the mid- and South Atlantic and Gulf coasts but had disappeared from most dune and beach areas long ago because of overgrazing by livestock.[8] Since it rarely, if ever, produces viable seed, it is slow to reinvade without the help of man or storm activity. Now that grazing has largely been eliminated, this dune grass is increasing throughout much of its range. It does have substantial value as a sand-stilling grass from about the mid-Atlantic coast to Mexico, as the initial stabilizer in the southern half of the region, and as a companion to other pioneer foredune plants elsewhere.[8,13] Characteristics of bitter panicum are

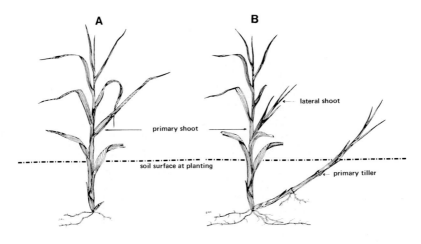

FIGURE 9.    Growth habit of bitter panicum. (A) Soon after planting, (B) after several months.

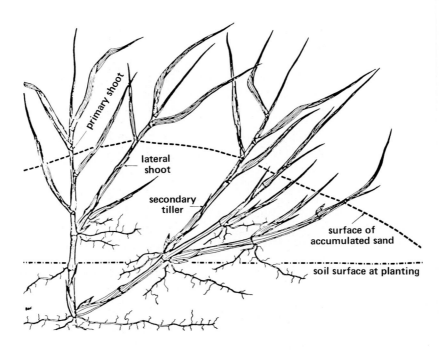

FIGURE 10.    Bitter panicum growth at end of first summer.

1.    Easy to multiply under nursery conditions but commercial availability is limited; the supply could be quickly expanded if demand develops.
2.    Rarely, if ever, produces viable seeds; therefore, must be propagated vegetatively.
3.    Free of serious pests under both nursery and dune conditions.
4.    Easy to harvest, store, and transplant. Survival of transplants is substantially superior to that of sea oats but inferior to American and European beachgrasses. The transplanting season is long.
5.    Growth is erratic the first year. Some transplants may start new growth immediately while others delay for some time.

FIGURE 11. Saltmeadow cordgrass starting a new dune.

6.     Sand trapping is efficient and the plants spread laterally in a way similar to sea oats. Dune slopes are more gentle.

7.     Cannot withstand as rapid a sand accumulation as American beachgrass or sea oats, and is particularly susceptible to burial during the establishment period.

8.     Wide variety of plant types are found in the field — creeping to erect, and delicate to robust. The best types for dune planting have not been determined and better propagation methods are needed.

### 5. Saltmeadow Cordgrass

This plant has smaller, more pliable leaves and stems than any of the other dunegrasses and offers less resistance to the wind; therefore, it is not as effective as a dune builder. However, saltmeadow cordgrass is very tolerant of salt and flooding and it seeds profusely, spreading readily by seeds. This grass is often among the first to invade low-exposed sites along the mid- and South Atlantic and parts of the Gulf coasts. It often initiates new dunes on low flats that may later become occupied by plant species better adapted to high dry conditions (Figure 11). It is widespread along the Atlantic and Gulf coasts. Characteristics of saltmeadow cordgrass are

1.     Easy to transplant where moisture is adequate but difficult on dry dunes. Plants are occasionally found on old high dunes where they apparently grew with the dune.

2.     Survival depends on type of planting stock. Plants must be young and uncrowded. Stock from old dune stands is usually unsatisfactory.

3.     Tolerant of reduced sand and nutrient supply. Persists longer under intermediate zone conditions than other pioneer zone species.

4.     Has a long planting season on suitable (moist) sites, winter through early summer.

### 6. American Dunegrass

This grass was apparently the only perennial dune grass occurring along the Pacific coast from British Columbia to southern California before the introduction of European beachgrass in the late 1800s. The same species also occurs around the Great Lakes and along the North Atlantic coast. Whether or not this is the same variety as the Pacific coast grass is not clear. Early workers[4] found European and American beachgrasses much easier to multiply and transplant. Consequently, little attention has been devoted to the propagation and use of American dunegrass until recently.[14] It has not been planted enough to know its role and value in building and stabilizing dunes. The grass invades foredune areas throughout most of its range and is capable of building foredunes. It has replaced European beachgrass on some older foredunes in the Pacific Northwest. American dunegrass seems to have a potential for more use as its requirements are better understood. Certainly, it seems to warrant further attention, particularly on the Pacific coast, where dune planting is presently done almost exclusively with exotic European beachgrass. Characteristics of American dunegrass are

1. More difficult to multiply than the beachgrasses, and is not available commercially.
2. Must be transplanted while dormant; therefore, has a shorter transplanting season than the beachgrasses.
3. Very temperature sensitive and should not be transplanted when temperature is likely to exceed 14°C.
4. Appears to be especially palatable to rabbits. This may account for poor survival in many cases.
5. Although the value of this grass for planting purposes is not clear, it may have a potential to reduce the hazards of the monoculture of European beachgrass. The advent of a serious pest in a monoculture could be catastrophic in many exposed sites.

## C. Propagation Techniques

Most of these plants produce viable seeds and at times spread into dune areas by seeds, but direct seeding is not usually a satisfactory means of establishing initial cover in the pioneer zone. In bare sand, seeds will too often become uncovered or buried too deeply before they can germinate and the seedlings become established. Consequently, with few exceptions, planting of foredunes is done vegetatively.

Adequate supplies of healthy planting stock are essential to any successful dune planting and the acquisition of plants is usually a major item of expense in planting projects. There are two principal sources: (1) nursery-grown plants, usually produced for this purpose from vegetative sprigs, but sometimes from seed and (2) plants obtained by thinning natural or cultivated established stands. One or two species lend themselves well to pot-grown seedling production.

Established stands are satisfactory and practical sources, particularly where small quantities are required. The best plants are usually found in back-dune and deflation plain areas where they are uncrowded and have not trapped large quantities of sand. The removal of planting stock from these stands reduces the cover only temporarily because the rhizomes left in the ground will usually revegetate early in the next growing season. Plants from foredunes are difficult to dig and make poor planting stock. Also, digging leaves the foredunes temporarily vulnerable to erosion.

Although established stands are often suitable sources of plants for small projects, the availability and harvesting expense usually dictate the need for nursery-grown material for large projects. This presents no particular problem with the more widely grown species if sufficient lead-time, 1 to 2 years, is allowed for plant producton. The market is usually quite erratic with surges in demand resulting largely from the occurrence of damaging storms and the initiation of large projects. Since it is difficult to hold planting stock in the nursery more than 1 or 2 years, commercial producers do not usually keep large stocks on hand. This makes planning essential to the successful planting of large projects.

### 1. American Beachgrass

This plant is relatively easy to produce under nursery conditions. It can be multiplied either vegetatively or by seeds, but the vegetative method is normally preferred (Figure

FIGURE 12. American beachgrass nursery midway in the first growing season.

12). Direct seeding is usually uneconomical because seed supplies are unreliable and weeds are difficult to control in seedling stands.

**Soil selection** — Any well-drained soil may be used and it is not necessary that the site be near the sea. Production of American beachgrass will be higher on the more productive soil and both planting and harvesting are easier on soils of a sandy nature (sands, sandy loams, and sandy clay loams) than on heavier soil. Since weed control often presents the greatest difficulty, emphasis should be placed on selecting relatively weed-free fields.

**Soil preparation** — A well-pulverized seedbed, suitable for the planting of normal field crops, is necessary to facilitate transplanting. Hard or cloddy seedbeds in heavy soils will lead to shallow planting and interfere with firming of the soil around the plants. At very weedy sites, fumigation with methyl bromide is advisable before planting. This requires a well-pulverized seedbed that is moist but not wet, and mild temperatures. Soils should be tested before planting and where nutrient levels are low, fertilizer applied.

**Transplanting** — Plants should be set in winter or early spring, one stem per hill, 45 to 60 cm apart in rows spaced 75 to 100 cm, depending upon available cultivation equipment. Planting should be to depths of 10 to 20 cm in moist soil and the soil pressed firmly around the base of the plants to avoid air pockets. Most mechanized transplanters, such as are used to set tomato, pepper, tobacco, etc. can be readily adapted to the planting of beach grass. Machine planting is preferable to hand planting under nursery conditions since it will ensure more uniform spacing for ease of cultivating.

**Culture** — Cultivation is necessary to control weeds but should be avoided where weed growth can be suppressed by other means. Fumigation with methyl bromide, where fully effective, should result in adequate weed control the first year. Most summer annual weeds can be controlled by the application of 1.5 kg/ha of Simazine® in the spring (follow timing directions for corn). Spray should be directed to the base of

FIGURE 13.    Planting stock of American beachgrass.

the plant to avoid the leaves. One or two nitrogen topdressings can be applied during the growing season on less fertile soils. Apply 40 to 60 kg of nitrogen per hectare and adjust the dosage to maintain healthy color and good growth. Excessive nitrogen application is wasteful and may be detrimental. Supplemental irrigation immediately following transplanting can be helpful to settle the soil around the plants and may increase production when used later in the season under very dry conditions. However, irrigation is not essential in the production of this plant since it tolerates long, dry periods under dune conditions.

**Harvesting** — Planting stock of American beachgrass may be harvested during the winter or early spring after one growing season. The individual clumps are loosened with a shovel or by a tree digger or plowing devices. They are then lifted by hand, shaken free of excess soil, and separated into individual "plants" of one to five stems. These may be transplanted immediately, stored for short periods by heeling-in out-of-doors, or held for a month or more in cold storage at about 0°C. Plants may be stacked upright in tubs or boxes for movement of short distances. However, where more handling and time may be required, package the plants in bundles of 500 to 1000 stems and wrap tightly in paper in a manner similar to that used for forest tree seedlings.

Avoid excessive drying of the basal part of the plant. Dipping the lower 10 to 13 cm in a clay slurry before packaging seals the base and gives a margin of safety during storage, transport, and planting.

Tops of bundled plants often require trimming to a length of 45 to 60 cm to facilitate machine planting and to reduce bulk and weight for handling and storage (Figure 13). However, there is some advantage in leaving plants untrimmed when hand planting criticial sites because the excess top growth will trap sand and protect the new plant during establishment.

Planting stock may be carried over in the nursery through the second and sometimes the third growing season with some increase in yield over the first year. However, after the third season, plants become too crowded, quality declines, and rodent infestation may become a problem, making it advisable to destroy the material and make a fresh start.

Normal harvesting operations leave large number of rhizomes behind. Consequently, if the field is left undisturbed, a dense volunteer crop of beachgrass will usually appear the next growing season. Such stands are capable of yielding substantial
numbers of plants if fertilized and kept free of weeds. However, harvesting of these
broadcast stands is more laborious than for row plantings and weed control is more
difficult. As harvesting is usually the most expensive step in the production of planting
stock, it is normally more economical to plant a new field each year than to attempt
to harvest volunteer plants.

### 2. Sea Oats

Propagation of this plant is different and more difficult than that of American
beachgrass. Sea oats will not thrive on sites very far from the dune environment, and
even under the best conditions, multiplication is slow. The reasons for this are not
clear, but it is known that in inland locations this grass falls prey to at least two pests:
a "helmethesporiumlike" leaf spot or rust and a stem borer, similar to the Hessian
fly, neither of which are serious pests in the dune environment. The leaf disease can
be controlled by carefully timed spraying with Daconil® or Bravo®; the stem borer
with a systemic insecticide. However, the most satisfactory nurseries have been within
a few hundred feet of the sea (e.g., Padre Island).

**Soil selection** — Sites in back-dune swale and deflation plain areas are preferred for
field-scale production of sea oats. Care should be taken to assure protection from sand
encroachment and stormtide damage on such sites. In the absence of such sites, sandy,
well-drained soils farther inland may be used, but disease and insect control will probably be required.

**Soil preparation** — A clean, weed-free area is essential for satisfactory nursery production. Foredune species do not compete well with weeds and the cost of hand weed-
ing of nursery areas soon becomes prohibitive. Sea oats are especially vulnerable to
competition because of slow initial growth. Since the more productive back-dune and
deflation plain planting sites are usually occupied by vegetation, eradication of the
growing plants and dormant seeds is the first step. This may be done on the clean sites
by cultivation, but heavily vegetated areas will require fumigation with methyl bromide.

**Planting** — Sea oats nursery areas may be established by transplanting wild plants,
direct seeding, or transplanting started seedlings in peat pots from a greenhouse. The
best wild plants are obtained from young clumps that have not yet become deeply
seated. These are usually found on young deflation plains or on wide areas in front of
the foredune (Figure 14). Clumps may be divided into single stems and set 15 to 20
cm deep in moist soil 45 to 60 cm apart. Rows should be spaced for mechanized cultivation.

Direct seeding is feasible when seed supplies are adequate and a high degree of weed
control is assured. However, sea oats seedlings start very slowly and are vulnerable to
insect and disease damage as well as to weed competition the first year. For this reason,
direct seeding is not recommended.

The initial period of slow growth can be circumvented by starting seedlings in peat
pots in the greenhouse during the winter and transplanting them to the nursery in early
spring. Although this step increases cost, it may save most of a whole growing season
in startup time. Seeds of the more northern populations of sea oats require a period
of about 30 days of cold storage to break dormancy.[15] Peat pot-grown seedlings may
be transplanted directly to the dunes but these tend to be more expensive than planting
stock from nursery beds.

**Management** — Plants will respond to small amounts of fertilizers (30 to 40 kg of
nitrogen, 10 to 20 kg of phosphate per hectare) on sites that are low in nutrients.

FIGURE 14.    Sea oats dunelets forming in front of a barrier dune, Hatteras Island, N.C.

Excessive fertilization will stimulate weed growth and should be avoided. Cultivation should be avoided except when required for weed control. Although some usable transplants are produced the first growing season, sea oats that remain undisturbed through the second growing season will result in more and stronger plants.

**Harvesting** — Sea oats are harvested by loosening the sand around individual clumps with a shovel or other tool, lifting the clump by hand and shaking it free of excess sand. It is usually not possible to pull unloosened plants without excessive damage. Clumps are then hand-separated into transplanting units of one or more healthy, vigorous stems. Transplants may be stored upright in tubs or baskets for handling or packaged in bundles as described for American beachgrass.

Care must be taken to avoid excessive drying of the base of the plant. Dipping the lower 10 to 13 cm in a clay slurry, immediately after processing, is suggested. When plants are not to be transplanted immediately, they may be held for periods of a week or so by heeling-in in moist sand. Sea oats plants do not store well in water.[8]

*3. Bitter Panicum*

This plant multiplies readily under nursery conditions, grows well on any well-drained soil, and has no serious pests. As with other dune grasses, nursery manipulation and harvesting are more convenient on sandy soils.

**Soil selection and preparation** — Any well-drained, sandy, weed-free soil is satisfactory. The site does not need to be near the sea. A well-pulverized seedbed should be prepared and, on very low fertility soils, a complete nitrogen, phosphate, and potash (NPK) fertilizer applied as for corn.

**Planting** — Wild stock may be planted. Seeds are not produced; therefore, only vegetative parts are planted. The best wild transplants come from backshore and young deflation plain areas and the back slope of foredunes where sand accumulation has been minimal. Plants are pulled by hand. They usually break off at the ground surface in the winter when brittle, but come up with roots and rhizomes attached during the growing season. Rooted and rootless stems (culms) are equally satisfactory for transplanting. From spring to autumn, two types of panicum stems are available: (1) primary stems from the previous year's growth which have flowered and are firm and brittle and (2) tillers which are actively growing stems. Tillers survive and grow best following transplanting during the growing season; primary stems are best when planting from autumn to spring.[8] Long primary stems may be divided to form 30- to 50-cm transplants. The upper and lower halves survive equally well. It is also possible to stretch planting stock supplies by planting large stems horizontally in furrows. Care must be taken to bury only to a depth of 10 to 15 cm and the stem tip should be left uncovered. This type of planting will result in a new plant at almost every node.

**Harvesting** — Planting stock may be harvested from nursery plantings after one growing season. Plants are pulled by hand or, in large-scale production, mowed and raked to obtain the top growth. The smaller stems, typical of old dense stands, do not make as satisfactory planting stock as the larger, more robust stems, typical of young, well-fertilized stands.

Pulled or mowed plants may be heeled-in in moist sand for short periods. This species may be stored up to a month by immersing the lower half in freshwater.[8] Plants may be stacked in tubs or baskets for transplanting. For more extensive handling, bundling and clay-dipping, as described for American beachgrass, is suggested.

Theoretically, a bitter panicum nursery can be left in place and harvested year after year if weeds are controlled. In practice, plants may become too crowded after 2 years or more and it is advisable to make a new start.

### 4. Saltmeadow Cordgrass

This is probably the most plentiful of the dune grasses along much of the Atlantic and Gulf coasts, occurring widely on low flats and deflation plains. However, it is difficult to obtain good planting stock from the wild. Stands on moist sites become dense and crowded, making harvesting difficult. Plants on dry infertile deflation plains lose vigor and survive poorly when transplanted. The best transplants are from rapidly growing, uncrowded young stands with relatively large stems. Plants are harvested by loosening soil around clumps with a shovel and hand lifting. Although very little nursery production of this plant has been done, it appears that there may be more to be gained by it than with any other dune species in terms of both transplant quality and economy. The production of pot-grown seedlings of saltmeadow cordgrass is relatively easy. Seeds should be stored dry and cold.

**Soil selection and preparation** — This grass may be grown on both inland and coastal sites. The prime requirement appears to be a clean, sandy loam soil with a moderately good moisture holding capacity. The seedbed should be well pulverized and if necessary, fumigated with methyl bromide.

**Planting** — Saltmeadow cordgrass can be grown from seed but for planting stock production, sprigging on 45- to 60-cm centers in rows 75 to 100 cm apart is best. Use 2- to 4-stem transplants from young vigorous stands, set 10 to 15 cm deep in moist soil. Although not essential, supplemental irrigation is often helpful since this species has a higher moisture requirement than the other dune grasses.

Excellent transplants of this species can be produced in the nursery in one growing season. They may be harvested by loosening individual clumps with a shovel or by lifting with a tree digger or similar tool. Clumps should be cut into smaller five- to eight-stem units for transplanting. The units may be stored temporarily by heeling-in in moist sand or stacked upright in tubs or baskets, preferably bundled and clay-dipped for handling and transport as described for American beachgrass.

On the basis of very limited experience, planting stock may be held over in the nursery for 1 or 2 years with no detrimental effect if plants do not become crowded and lose vigor. In that case, it is best to start a new planting.

### 5. European Beachgrass

This is the easiest of the foredune species to propagate and can be produced most efficiently in nurseries. However, much of the planting stock for large stabilization projects in the Pacific Northwest came from the extensive established stands in that region.

**Soil selection** — Any sandy, well-drained soil will serve. Much of the production has been on dune sand, but it is better that the nursery site not be exposed to substantial sand movement. The more protected sites should be weed-free. Because this plant is temperature sensitive, nursery production probably should be kept well within the modifying influence of the water.

**Soil preparation** — Little preparation is required in dune sand except the removal of existing vegetation. Fumigation with methyl bromide is advisable for weed control. Soil should be tested and nutrient deficiencies corrected.

**Planting** — European beachgrass may be seeded in the nursery but in view of seed costs and availability, transplanting is more practical. Nurseries should be stocked during the winter or early spring when the air temperature is likely to remain at or below 16°C for several days and the soil is moist. Plantings spaced 30 to 45 cm apart in rows, one stem per hill, is best for limited supplies of planting stock. However, where plants are available, closer spacing will greatly increase production per hectare. Brown and Hafenrichter[16] obtained the highest production from five stems per hill spaced at 30 by 30 cm. However, they were working on dune sand under exposed conditions which probably placed a premium on early stabilization. Machine planting is preferred and plants should be set 15 to 30 cm deep.

**Culture** — Cultivation should be avoided except where essential for weed control. Production will usually be increased by the application of 40 to 60 kg of nitrogen per hectare soon after spring growth begins. Fertilization is essential for strong nursery stock on sands and infertile soils.

**Harvesting** — Plants in nurseries may be pulled by hand with little or no digging. Plants from dune stands are harvested by cutting them about 5 to 8 cm below the surface to leave one or two underground nodes, using a sharp, flat-blade garden spade with a straight cutting edge, 15 to 20 cm wide. Clumps are lifted, shaken free of excess soil, cleaned of trash, the underground stems broken back to one or two nodes, and separated into individual one- to five-stem transplants. These may be heeled-in narrow trenches or stacked upright in tubs or boxes for local handling or packaged for shipping or long-term storage in bundles of 100 to 1000 plants with the lower half wrapped in paper to avoid drying, as is done with forest seedlings. Dipping the lower 8 to 10 cm in a clay slurry before packaging provides an economical deterrent to drying and a desirable margin of safety against careless handling and planting.

No data are available on suitable temperatures for cold storage of European beachgrass. The optimum temperature, 0°C, for American beachgrass is probably satisfactory for this species.

Stems should be trimmed to an overall length of about 50 cm after packaging for easy handling, storage, and transplanting.

### 6. American Dunegrass

This species is considerably more difficult to propagate than the beachgrasses[4,6] so attention has been focused laregly on the beachgrasses and little specific information is available on the culture of this dunegrass. Therefore, the following suggestions are speculative.

**Soil selection** — As the grass thrives under foredune conditions, nurseries probably should be located on very sandy soil close to the sea. Sites should be weed-free since this species is not as competitive as the beachgrasses. Fumigation should be used for weed control.

**Soil preparation** — Little or no preparation is required in sand. Tillage to loosen hard-packed sites may be needed to facilitate transplanting.

**Planting** — American dunegrass should be transplanted in the same way as American beachgrass. Plant 15 to 20 cm deep and 45 to 60 cm apart in rows 75 cm to 1 m apart. Since planting material is not likely to be plentiful and the plant spreads, planting one stem per hill is advisable. The transplanting of dormant plants appears to be the key to survival for this species. The dormant period in the Pacific Northwest extends from late November to the end of February. It may not go dormant south of this region. The temperature probably should be below 14°C at planting and for several days thereafter.

**Culture** — This plant thrives under the rapid sand accretion and, therefore, probably responds to fertilization. Suggested application rate is 40 to 80 kg of nitrogen per hectare from a soluble source in early spring or as soon as new growth begins.

**Harvesting** — Harvest only during the dormant season. Plants may be loosened with a shovel, lifted, shaken free of sand and dead trash, and broken into one- to five-stem plants. Cutting of stock a few centimeters below the surface with a sharp shovel may be necessary if sand has been deposited during the current growing season. Plants may be stored during cold wather by heeling-in. They may be stacked in baskets, boxes or tubs, or clay-dipped and packaged for handling as described for beachgrass. Trim tops to 50 cm for easy storage, handling, and transplanting. Take extra care to avoid drying and overheating. Planting should be as prompt as possible after digging.

## III. RESTORATION TECHNIQUES

### A. Soil Moisture

Water is essential to the establishment of plants. The low water holding capacity of sand can cause serious failure of plantings. However, compensating factors in the beach and dune system often make it possible to work around this problem. On low-lying beaches in the Atlantic and Gulf coast barrier islands, the water table is always close to the surface. Dahl et al.[8] found that moisture on the backshore (elevation, 1.3 to 1.6 m mean sea level [MSL]) was adequate at a depth of 15 cm except during extreme droughts. Moisture content was usually at or above field capacity, the water table was usually within 60 m of the surface, and capillary action kept the sand moist. These sands drain excess water readily and surface drying is extremely fast but total water loss is low. The layer of dry sand minimizes evaporation losses from below the surface as long as the layer remains. Consequently, even on elevated surfaces such as fence-built dunes where the water table is not near the surface, the sand may remain moist a few centimeters below the surface for considerable periods of time. Dune plants have various specialized adaptations for surviving substantial periods of low moisture. The least tolerant stage lies between transplanting and the development of a new root system that is adequate to extract moisture throughout a large volume of sand. Consequently, transplants must be set deep enough for root development to occur before complete drying. On most U.S. coasts, the sands are usually moist during most of the recommended transplanting season; lack of moisture is a problem only in unusually dry seasons.

Moisture content is more important in the Gulf coast where temperatures remain relatively high throughout most of the year, rainfall is erratic, and evaporation rate exceeds precipitation. Dahl et al.[8] found it necessary to irrigate fence-built dunes and other elevated areas behind the backshore before planting, primarily to firm the sand, except during and soon after heavy rains. They used sprinklers supplied by water pumped from open-pit wells dug immediately behind the dune line. Water was applied at the rate of about 0.5 cm/hr; it took about a day to wet the sand to a depth of 15 cm. The ground water was excessively saline at times but up to about 3% was tolerated by the plants.

Irrigation after planting is not generally worthwhile. Sand does not dry rapidly below the 15-cm level and irrigation does not raise the water table. Consequently, if the planting zone is moist from rainfall or preplanting irrigation, further watering adds little. Small-scale plantings where intensive management is possible may justify irrigation of the planting. Irrigation may be used under extreme drought conditions and to leach out salt following inundation by saltwater. However, the plants that are well-adapted to the dune and beach environment and have adequate moisture at planting, usually grow and survive with little added help.

Irrigation is sometimes the only way to firm the surface and prevent dry sand from refilling holes or furrows before plants can be inserted. Hand planting can tolerate less than 1 5-cm depth of dry sand but a mechanical transplanter can operate through a layer twice as thick. When the dry layer exceeds these limits, the only alternatives are to irrigate or wait for rain.

## B. Salinity

Salinity is a potential inhibitor to plant growth along any seashore. Salt is deposited on beaches and dunes in substantial quantities by salt spray or by flooding.

Fortunately, the potential of salt damage to the establishment and growth of dune plants is greatly tempered by the rapid leaching of the dune sands. These sands have almost no retentive capacity for salt and only a small amount of rainfall is required to remove salt from the plant zone. Dune plants can tolerate moderate concentrations of salt and some of them do not absorb salt through their leaves. All can tolerate some salt in their root zone. In the upper beach and coastal dunes, the lighter freshwater tends to float on top of the heavier seawater. In humid climates, percolating rainwater causes the development of lenses of fresh to mildly brackish water on top of the salt-water under even the smallest dunes[17] and may create substantial reserves of freshwater under larger dune systems. This resistance to mixing may also allow plants an escape from salt damage following flooding. If the sand is moist, the seawater will drain off with little infiltration into the freshwater and little or no effect on the plants, except in low spots where it becomes trapped and evaporates, leaving a high concentration of salt.

Another possible mechanism in reducing the salt effect of inundation is the rise of the water table in the beach and dune area that accompanies a general rise in the tide level. The freshwater lense tends to rise with the water table leaching salt from the root zone.

Salinity is not usually a major barrier to the establishment of adapted foredune species on most U.S. coasts, but it can become a serious problem in the Gulf coast where low sites have to be planted under conditions of warm to hot temperatures, low and erratic rainfall, and high evaporation rates. Dahl et al.[8] observed this in planting hurricane surge washovers on South Padre Island. They constructed a broad flat dune across a washover pass by trapping sand with 60-cm sand fences to leach out salt so bitter panicum and sea oats could survive. This lowered the subsurface salinity from 15% to near zero in 1 year.

Dahl et al.[8] also planted exposed back beaches and protected areas on North and South Padre Islands to study the effect of inundation on survival of transplants. Results were erratic — some plantings were essentially eliminated by storm surges and some were little affected. In general, established plantings were more tolerant to exposure but the increased survival on protected sites could not be attributed to only salinity differences. Drifting sand appeared to be equally important in several instances.

## C. Fertilization

Dune and beach sands undergo extensive leaching during formation, transport, and deposition. Consequently, they are inherently low in most nutrients essential to the growth of higher plants. In nature, most dune plants persist under a chronic deficiency of these nutrients. While typical dune species are well-adapted to a low nutrient regime, most respond noticeably to fertilizers. This does not mean that the general use of fertilizers in the dune and beach system is desirable, but that fertilization can be a useful management tool.

Fertilization is useful for rather definite and restricted purposes: to speed up the establishment of new plants, increase growth, and increase sand-trapping capacity, and thus improve their chance of survival; and to revive declining stands to maintain protective cover in areas receiving a reduced or intermittent sand supply.

Response to fertilizers is usually most pronounced on old, leached sands in backdune and deflation plain areas that are cut off from fresh sand supplies. Response is likely to be much less on sites with active sand accretion. However, the initial establishment period is usually the most critical for dune plantings and even a moderate acceleration of growth at this point may mean the difference between success and failure, particularly on the more exposed sites. Fertilization, by reducing the limitations imposed by nutrient supply, enables better plant growth during favorable periods. Fertilization is usually discontinued in the active sand zone as soon as plants are established.

Most pioneer species, especially the major dune builders, lose vigor very rapidly when they no longer receive fresh sand. They may die out altogether and be replaced by plants of the intermediate zone that are much less effective in sand trapping and stabilization. This is the normal succession in this situation and unless there are practical reasons to do so, there is no point in interfering with it. However, the effect of fresh sand, or the lack of it, on pioneer dune plants is in part, at least, a nutrient effect and it is possible to revive sand-starved stands through fertilization. This is very useful whenever it becomes necessary to restore or maintain a vigorous foredune-type cover on areas that become cut off from fresh sand. Specific suggestions for fertilizer use are presented for the geographic regions later in this section.

Suggested fertilizer schedules are based on the use of standard commercial nitrogen and phosphate materials or mixed fertilizers of nitrogen and phosphate. Conventional soluble sources are surprisingly effective in light of the inability of dune sand to retain nutrients. This is apparently due to the interception of nutrients as they leach downward through the extensive root systems of dune grasses. Response is largely to nitrogen and sometimes to phosphorous. Consequently, a ratio of 3 parts nitrogen to 1 part of phosphate makes a good dune fertilizer. Occasionally, it may be more convenient to use commercially available fertilizers containing potassium, in addition to nitrogen and phosphorus. The potassium will do no harm but observable response to potassium or micronutrients is unlikely as long as the dune grasses are subjected to salt spray which supplies them. Where the suggested fertilizers are not readily available, waste can be minimized by alternating one application of 8-8-8 (8% N, 8% $P_2O_5$, 8% $K_2O$) or 10-10-10, for example, with two or three applications of a straight nitrogen source, such as ammonium nitrate, to approximate the suggested amounts of nitrogen and phosphate. Slow-release materials, particularly those containing slowly soluble nitrogen, reduce both leaching losses and the number of applications needed; however, results with these have been inconclusive. Most slow-release fertilizers are not fully effective unless placed deep enough to remain moist most of the time. Some slow-release materials have been used to speed establishment of American beachgrass by placing a small amount in each planting hole at time of planting.Although this promotes rapid growth for a year or two, it may overstimulate the plants, causing extensive die-out later, particularly if fertilization is not maintained. All slow-release materials are considerably more expensive than conventional sources.

Conventional fertilizer materials should be broadcast by ground or aerial equipment, and always pelleted or granulated to minimize drift. A helicopter is particularly well-suited, provided the area involved is large enough to warrant its use. The advantages of using a helicopter are better distribution of the fertilizer, no wheel-track damage to dune cover, a good distribution of the pellets of granules under windy conditions because the down blast from the rotor prevents pellet drift, and helicopter landing requirements permit loading close to the area to be fertilized.

### D. Planting
#### 1. American Beachgrass
This grass is native to the North Atlantic coast and the only plant regularly used as the initial stabilizer in the foredune zone there. It is easy to multiply and transplant under nursery conditions, grows rapidly when transplanted, and is exceptionally effective in trapping sand and stabilizing dunes. It is usually planted in pure stands which makes the planted areas subject to rapid deterioration in the event of serious pest damage.

Recently, severe losses of American beachgrass stands in North Carolina have been caused by infestations of a soft scale, *Eriococcus carolinae*.[11] This pest appears to be widely distributed along the mid-Atlantic coast and probably occurs on the North Atlantic coast. This plant is very susceptible to Marasmius blight[12] in the warmer part of its range. For this reason, immune species should be interplanted with American beachgrass where possible to reduce this hazard.

#### a. Planting Methods
Planting is done by hand on small areas and rough or steep terrain, and by machine on large, smooth sites. In hand planting, plants are inserted in individual holes opened with a shovel, spade, or dibble. This is best done by two-man teams, one opening the hole while the other inserts the plant and firms the sand around it. Machine planting is done with tractor-drawn transplanters designed to set crop plants such as tobacco, tomato, cabbage, etc. Most machines can be readily adapted to transplanting beachgrass by extending the openers or shoes to provide a deeper furrow in which to set the plant. Both one- and two-row machines are used. Wheel tractors are faster on smooth, relatively level sites; crawlers are needed on rougher sites.

#### b. Depth
American beachgrass should be planted 20 to 25 cm deep or deeper (30 to 35 cm) in loose, dry sand. The plants must be set deep enough for the basal parts to remain in moist sand until new roots develop to anchor them and new top growth can emerge to trap sand. The deeper they are placed in the moist sand, the less chance of being blown out before becoming established. Shallow planting is the most common cause of failure. It is difficult to open holes or furrows to the proper depth in hard-packed sand and it is more difficult to keep them open long enough to insert plants through a thick layer of dry, loose sand. This problem can usually be overcome by using more power, but if the sand is dry, it may be necessary to irrigate or wait for rain.

#### c. Planting Date
This plant transplants exceptionally well and can be transplanted satisfactorily when dormant or growing. It has a long transplanting season — successfully from October through May with the preferred period running from February through April.[5,18,19]

#### d. Planting Stock
Transplants should have one to several healthy, vigorous stems (culms) (Figure 13). Multiple stems planted in the same hill need not be attached to each other. Larger plants are preferred because the first year growth is definitely related to the number

of stems planted per unit area. Consequently, on critical sites where rapid stabilization may be essential, five or more stems per hill are suggested. However, normally spaced plantings of one stem per hill will cover well in the first growing season and there is little difference in multiple-stem plantings the second year. Planting stock represents a significant part of the total cost of a planting so one to three stems per hill are usually planted on all but the most critical sites. The critical sites are the windward slopes of large, mobile dune areas that receive unusually large volumes of sand, blowouts in or between dunes, and areas vulnerable to storm waves.

### e. Spacing

The exact spacing and pattern is important in the design of a dune grass planting. Spacing that is too wide will usually result in partial or total failure; spacing too close is wasteful. Planting costs are roughly proportional to the number of hills planted. For example, a 30-cm spacing requires four times as many hills per unit area as 60-cm spacing and costs about four times as much.

The spacing and pattern should be determined by the characteristics of the site and the objective of the planting. A strip of American beachgrass, 8 to 12 m wide, planted 45 cm on centers will, with normal development, effectively stop the movement of windblown sand in the last half of the first growing season. Small blowout areas should be planted at a spacing of 45 cm or less. Stabilization of a large area of bare sand will require a spacing of 45 to 60 cm. In more protected sites, 1 m spacings are adequate.

### f. Fertilization and Management

American beachgrass responds well to the addition of nutrients and the judicious use of fertilizers is useful in the management of this plant. Plant response varies widely — it is least under the condition of rapid sand accumulation and greatest on old, leached sands in back dunes and deflation plains. Growth on sites that do not receive fresh sand may be increased up to tenfold by fertilization. Response is chiefly to nitrogen and occasionally to phosphorus. Fertilization is used primarily for two purposes: (1) during establishment, to improve survival and (2) to maintain a vigorous protective cover on areas that do not receive sufficient fresh sand.

New stands will often benefit from the application of 100 to 150 kg of nitrogen per hectare and 30 to 50 kg of phosphate per hectare the first growing season. Application should begin as soon as new growth emerges and the total amount for the year should be divided into two or three applications, spaced 4 to 6 weeks apart. Fertilization after the first year should be adjusted to growth and appearance of the plants. It is usually not needed with substantial sand accumulation. Excessive fertilization is wasteful and can be harmful.

The same general principles apply to sand-starved stands. Poor stands will benefit from up to 150 kg of nitrogen and 30 to 50 kg of phosphate per hectare annually in two or three applications each year for 1 to 3 years. Stands that have not seriously declined may be maintained with an application of 30 to 50 kg of nitrogen per hectare applied in early spring, at intervals of one or more years. Fertilization practice should always be adjusted to the growth and appearance of the grass. Excessive growth may mat down and promote disease damage and plant loss.

Fertilizers containing nitrogen and phosphate in the suggested ratio are not widely available and, lacking these, the same effect can be obtained by alternating application of 10-10-10 or the equivalent with one or two applications of a straight nitrogen material such as ammonium nitrate, ammonium sulfate, or urea to approximate the desired ratio of nitrogen to phosphate. The added potassium is of no value but not harmful.

## 2. Bitter Panicum

This plant is increasing along the South Atlantic coast. It seldom produces viable seeds, so spread into new areas is slow and sporadic. It is readily propagated vegetatively.

It has promise as a companion to American beachgrass and sea oats along the Carolina coasts and an initial stabilizer farther south.

### a. Planting Methods

Transplanting is the same as for American beachgrass. Stems (culms) are set upright by hand or by tractor-drawn mechanical transplanters.

The ease with which bitter panicum forms new tillers at buried nodes (joints) permits modifications in this planting procedure. Long stems may be planted horizontally in furrows, 10 to 15 cm deep, with the tip of the stem left exposed. This results in a new plant at nearly every node. This method has been successful in nurseries to increase planting stock supplies. The method results in a new plant for every 15 to 20 cm compared with one per 30 to 60 cm where stems are set upright in normal fashion. The method is difficult under beach and dune conditions where depth control is a problem.

An alternate method which has been successful requires mowing and raking top growth from the nursery, broadcasting this material on the sand surface, and covering it by discing or plowing. This approach is more mechanized and may result in denser stands than either type of row planting. It is also cheaper where planting stock is plentiful and close at hand. Again, the principal problem is control of planting depth. Shallow planting causes the new shoots to die during dry periods before becoming fully established; planting too deep will cause stored food reserves to become exhausted before new shoots can reach the surface.

Neither method is reliable where substantial sand movement, either erosion or accretion, occurs during the establishment period.

### b. Depth

Planting depths for bitter panicum are similar to American beachgrass. Plants should be set 20 to 30 cm deep to prevent drying and subsequent loss that will lead to blowouts. This grass is more sensitive to sand burial than American beachgrass. Dahl et al.[8] reported low survival of bitter panicum when sand accumulation buried the uppermost living part of each transplant more than 15 cm deep. Survival of unburied plants was 68%; survival of plants buried 15 cm deep was 29%. This illustrates why control of planting depth is critical in furrow planting and plowing or discing of bitter panicum stems. It also explains the poor planting results near unfilled sand fences where sand accumulates rapidly. Care is required to avoid burying the grass too deep. This is another reason to plant sea oats, American beachgrass, or both with this grass.

### c. Planting Date

Bitter panicum is similar to sea oats in response to planting dates and conditions. It can be transplanted with some success about anytime of the year, provided moisture conditions at and immediately following transplanting are favorable. Dahl et al.[8] obtained excellent survival in each month of the year in at least 1 out of 5 years in the Gulf region. They found that survival of plants transplanted in the summer was higher than those planted in winter. They considered autumn the least desirable time to transplant panicum because transplants remained dormant over winter and did not resume growth until late spring. In the meantime, the planting was exposed to the hazards of winter weather — blowout, burial, and saltwater inundation. They concluded that as long as environmental conditions (principally moisture) were favorable, the best planting time for bitter panicum is in late winter or early summer. Experience along the South Atlantic coast agrees with this conclusion except a later (March) starting date is necessary in the northern half of the region.

FIGURE 15.    Primary stems and tillers of bitter panicum.

### d. Planting Stock

Two distinct types of bitter panicum stems (primary stems and tillers) are available in the autumn and again in the spring and early summer. Primary stems represent mature growth from the previous year which has flowered and is generally dry and brittle. Such stems may be 1 m or more in length and most of the lower leaves are dead with the terminal leaves still green. These are preferred for transplanting in autumn and early spring. Tillers are young, growing, succulent stems with green leaves and are usually smaller than primary stems (Figure 15). Actively growing tillers are the best planting stock in late spring and summer when they become established and grow quickly. Consequently, planting stock should be according to the season.

Size of primary stems is important in selecting planting stock. Dahl et al.[8] obtained higher survival, threefold to tenfold, from large stems (60 cm long), as compared with small primary stems (30 cm long). Small primary stems are usually found in old, crowded, or starved stands; their poor performance is probably due to their limited reserves of stored food. Use of crowded nursery plants is not economical as it requires a high density planting to make up the higher mortality.

In upright planting, stems longer than 50 to 60 cm should not be used although primary stems from vigorous nursery stock are often 1 m or more long. Long stems may be divided into two or more pieces for planting with little difference in survival between top and bottom segments.[8] Consequently, where large primary stems are harvested, dividing them will double the number of usable transplants without reducing survival and will decrease the cost.

### e. Spacing

The spacing and planting pattern for bitter panicum is the same as outlined for American beachgrass. Excessively wide spacing invites failure; dense spacing is wasteful. The planting pattern should take ito account the site and the objectives. In the more southerly part of this region, a planting of bitter panicum, 15 m wide and 45 to 60 cm on centers, will effectively stop movement of windblown sand across the strip by the end of the first growing season. This spacing should also be used where a dense spacing is needed as on blowouts, diseased spots, or large bare dunes.

The shorter growing season in the northern part of the region (coast of the Carolinas) would require closer spacing if this species was planted alone. However, it is usually cheaper and more effective to use it in a mixture with American beachgrass, with or without sea oats. In this case, spacing is the same as that used farther south, 45 to 60 cm on centers in uniformly spaced plantings or at appropriate spacings for graduated patterns.

One stem of bitter panicum is usually planted per hill. Spacing and planting pattern suggestions are based on this. Plantings of multiple stems per hill are not warranted except where planting stock is limited to very small tillers.

### f. Fertilization and Management

Bitter panicum responds to fertilizers in the same general way as American beachgrass. Fertilizer will speed up plant establishment, help maintain a vigorous protective cover on areas receiving limited amounts of fresh sand, or revive and maintain vigor of old sand-starved stands. Rates of application are the same as for American beachgrass, ranging from two to three applications per year of 40 to 50 kg of nitrogen per hectare and 10 to 15 kg of phosphate per hectare to establish stands or reactivate old ones, to a single similar application at intervals of one to several years to maintain vigor. Fertilization should be adjusted to the growth and appearance of the grass.

Fertilization suggestions are in terms of standard commercial nitrogen and phosphate fertilizer materials. The use of some of the recent types of slow-release materials is justified in some cases but care should be taken to be sure that the type is effective under dune conditions. Fertilizers that require fairly constant moisture or certain microorganisms to be available to plants are not effective in the dunes.

Management of bitter panicum stands after reaching full cover is similar to that for American beachgrass. The stand requires reasonable protection from foot and vehicular traffic, and replanting of breaks that might lead to blowouts. This grass is very palatable to livestock[8] and other grazing animals such as rabbits, and may require protection.

### 3. Sea Oats

This grass dominates foredunes from Cape Hatteras south to Mexico. It is more persistent than other foredune species in the back-dune areas of this region, but it is not a good initial stabilizer. It grows slowly, is difficult to propagate, and is not widely available commercially. Consequently, it should not be planted in pure stands. Because sea oats is an excellent sand trapper, well-adapted to this region, and very persistent, it is useful to include it as a minor part of a planting mixture. It can be mixed with either American beachgrass or bitter panicum along the coast of the Carolinas, and with bitter panicum farther south. Sea oats will spread as the other grasses die, thin out, or are overcome by excessive sand deposition.

### a. Planting Methods

Planting is done with the same equipment as American beachgrass. Transplants tend to be more variable in size and are slightly more difficult to machine plant than American beachgrass plants.

### b. Depth

Planting depth is the same as for American beachgrass, 20 to 30 cm. However, sea oats is a slow starter and it is essential that it be set to the full depth to prevent drying before establishment and to avoid blowouts.

### c. Planting Date

Results of planting trials have been quite variable. Moisture conditions at and immediately following planting are more critical in the survival of sea oats than the season

FIGURE 16.    Large to small sea oats plants.

of the year or physiological condition of the plant. Transplanting is probably successful in any month of the year in the southern part of the region. Dahl et al.[8] obtained
satisfactory survival in July and September during a wet summer. However, after 5
years they concluded that January and February were usually the best months. In the
more severe climate of the South Atlantic coast, February to April appears to be the
optimum time; later planting is feasible only under very favorable moisture conditions.

### d. Planting Stock

The number of stems (culms) per transplant is not a factor in the survival and growth
of sea oats. Dahl et al.[8] found little difference between one, three, and six stems per
hill. Single-stem transplants survived well under favorable conditions; under poor conditions survival of three or six stems was slightly better. Multistem transplants do not
appear to be justified.

The range in stem size is greater in sea oats than in American beachgrass (Figure
16). Small plants survive poorly. Dahl et al.[8] obtained substantially better survival
from medium to large stems (75 cm to 1.5 m tall) as compared with very small to small
(45 to 75 cm tall) stems. However, they concluded that the difficuty in digging and
processing very large plants negates the survival advantage. Dahl et al.[8] concluded that
there is an advantage in using 2-year-old nursery-grown sea oats as this will furnish a
large proportion of intermediate-size plants.

### e. Spacing

The slow starting of sea oats makes it necessary to use a denser spacing than is used
with American beachgrass. However, sea oats is not usually planted in pure stands
because of the high cost of the stock and because sand-trapping capacity and stabilization can be developed quicker by including a high proportion of one or two less
expensive species in the planting. Sea oats plants become unusually effective as sand
trappers once they become well established. Dahl et al.[8] described a sea oats strip,
spaced 60 by 120 cm, 15 m wide, planted in April 1972, with only 36% survival, that
effectively stopped sand moving across it by the end of the first growing season. The
strip had built a dune over 2 m high by March 1974. In view of this, it is doubtful
that spacing closer than 50 cm on centers would be economically justified. Sea oats is
usually planted as a minor component of a dune grass mixture. One or two rows are
generally included in barrier dune plantings or in every 10th to 20th row in very large
plantings.

Direct seeding is not practical to establish sea oats on bare dunes. Seeds can be used to introduce the grass into new plantings of other species such as American beachgrass and bitter panicum. Seed heads can be gathered when mature (October in the Carolinas or September in Florida) and broadcast over the new planting where they will be trapped, covered by sand, and germinate later.

### f. Fertilization and Management

Sea oats respond to the addition of nutrients in the same way as American beachgrass. A moderate application of nitrogen and phosphate can be used to speed establishment of new plantings and to maintain growth and vigor in sand-starved areas. However, this plant is considerably more tolerant of low nutrient levels than American beachgrass and will persist in back dunes and deflation plains for long periods without the addition of nutrients or fresh sand.

### 4. Saltmeadow Cordgrass

This plant grows abundantly along the Atlantic coast. It has not been planted extensively for dune building and stabilization because it is not as effective a sand trapper as the other dune grasses. However, it frequently initiates and builds low dunes which may later be taken over by other plants (Figure 11). It is more salt tolerant, but less drought tolerant than American beachgrass, bitter panicum, and sea oats. It is particularly well suited for planting on low, moist sites where periodic salt buildup occurs. It probably has greater utility for initial stabilization of this type of site than has been generally recognized.

### a. Planting Methods

Saltmeadow cordgrass is planted the same way as the dune grasses. The finer, more pliable stems and the need for multiple-stem transplants make it more difficult to machine plant than the other grasses.

### b. Depth

Planting depth on drier sites is the same as for American beachgrass, bitter panicum, and sea oats. Saltmeadow cordgrass should be set 15 to 20 cm deep to keep it in the moist zone. On low-lying moist sites, planting depth may be reduced to about 15 cm.

### c. Planting Date

Little information is available on the planting date for saltmeadow cordgrass, but it appears to behave somewhat like bitter panicum. The best planting period is probably late winter and spring and it can be transplanted into the summer if moisture conditions are favorable.

### d. Planting Stock

The nature and condition of the planting stock appears to be a major factor in the survival of plantings of saltmeadow cordgrass. Stock must come from vigorous, uncrowded stands. As plants become crowded or starved, their value as transplants declines drastically. Suitable material is difficult to obtain from the wild; therefore, nursery production is necessary. The stems are small and multistem transplants are highly desirable. The number of stems to be used will vary with stem size and stored food content but generally there should be five to ten stems per transplant. These plants are usually too long to machine plant without trimming. Also, due to the larger number of more pliant stems and leaves, more care in trimming is required to avoid problems in feeding through the planter.

### e. Spacing

Spacing should be adjusted to the nature of the site and the objective of the planting. The plant is a less effective sand trapper than the grasses with larger, stiffer stems and leaves which suggests closer spacing. With vigorous plants, adequate nutrients, and favorable moisure, it is quick to establish and cover over. Consequently, spacing of 40 to 60 cm on centers is probably adequate for single-species plantings on suitable sites. Where a dune ridge is to be built, it should be planted on the same graduated planting pattern described for American beachgrass.

### f. Fertilization and Management

This plant responds well to nutrient supply. Fertilization is probably the key factor in the success of healthy planting stock. Fertilization should be adjusted to growth and appearance of the grass, but it will usually benefit from 100 to 150 kg of nitrogen per hectare divided into two or three applications the first year. After that, fertilization can be reduced to a single application for 1 or 2 years, then discontinued until the stand appears to need additional nutrients.

Saltmeadow cordgrass requires protection from excessive traffic but vigorous stands are considerably more tolerant than most dune plants. It will also tolerate a moderate amount of mowing. This was the major salt-hay species harvested in the past in substantial quantities along the North Atlantic coast.

### 5. European Beachgrass

This is essentially the only species planted along the Pacific coast since the 1930s for the initial stabilization of blowing sand. The ease with which this introduced beach-grass can be increased and transplanted is a great advantage in stabilizing large areas. No other plant is as inexpensive to use. In smaller plantings other species such as American beachgrass might offer advantages that could outweigh their higher costs. European beachgrass is a very effective sand trapper (Figure 17). Unfortunately, it forms dense stands but lateral spread is slow which results in dunes with steep windward slopes. Good initial stands and regular maintenance of this species are essential as it does not spread and fill-in as well as most dune grasses. European beachgrass is competitive in the active sand zone and excludes native species. This gives the dune a monotonous appearance and makes it difficult to establish mixed plantings. Behind the primary zone, the grass loses vigor and declines rapidly as its sand supply is cut off making it essential to introduce species that can take over and replace it.

### a. Planting Method

Planting is done by hand on small areas and steep slopes and by machine on large, smoother sites. In hand planting, it is best to work in two-man teams with one man opening the hole with a spade, shovel, or dibble and the other inserting the plant and firming the sand around it. Machine planting is done with tractor-drawn transplanters built or adapted for this purposes. With either method, sand must be firmed around the base of the plant to exclude air pockets.

Planting should not be done when the temperature exceeds 16° C or is at or below freezing. Moist sand should be within 8 cm of the surface.

### b. Depth

Suggested minimum planting depth on drifting sand is 30 cm. This depth may be impractical in the hard-packed sand of some foredune sites. It is essential that plants be set deep enough to remain in moist sand during establishment and to anchor them against strong winds during the planting season.

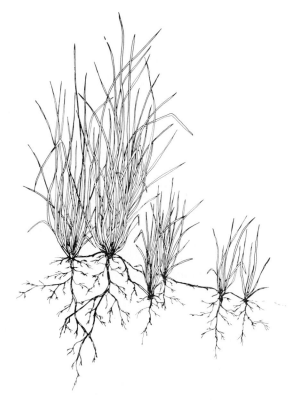

FIGURE 17.    Growth habit of European beachgrass.

### c. Planting Date

Suitable temperature and moisture conditions are more important to the survival of this grass than the planting date. These conditions are usually optimum in this region during the late autumn, winter, and early spring months.

### d. Planting Stock

Plants should be shaken free of sand, separated, cleaned of dead stems and trash, and pruned to an overall length of 50 cm. Stock may be temporarily stored by heeling-in in narrow trenches if soil is well-drained. For long-term storage, clay-dip, bundle, and hold at about 1° C. Three to five stems per hill are usually planted because dense stands are essential under the wind conditions in this region. Anything with less than a 90% survival rate should be replanted.

### e. Spacing

This is a critical factor in determining both the probability of success and the cost of a planting. Planting costs are proportional to the number of hills planted. Spacing and planting pattern should be adapted to the site and the result desired. Generally, a 45- by 45-cm planting with three to five stems per hill is sufficient on all but the more critical sites such as steep windward slopes and the tops of foredunes. A graduated planting pattern[20] is better for building a foredune by allowing the center ridge to develop rapidly and avoid the steep seaward slope typical of this species. A pattern of several rows with plants spaced 30 by 30 cm on centers and the other rows forward and backward of the center of the strip with plant spacing graduated from 45 by 45, 60 by 60, and 90 by 90 cm will build a more stable foredune at less cost than a uniformly spaced planting. Total width of the planted strip (20 to 40 m) will depend largely upon the volume of moving sand anticipated.

FIGURE 18.    Morphological features of American dunegrass.

#### f. Fertilization and Management

Fertilization is extremely important to the establishment of European beachgrass back of the foredune. It is probably less critical within the foredune but because of high-energy conditions there stimulation by fertilization is desirable. Response is limited to nitrogen, with 40 to 60 kg of nitrogen per hectare applied in early April or when rapid spring growth begins. Ammonium sulfate is the customary nitrogen source because it is readily available. Other standard sources are probably as satisfactory.[5]

A response to phosphorus could occur on some sites. Andriani and Terwindt[7] found no benefit from phosphates applied on the well-washed sands of dikes built during the construction of Europoort. European beachgrass may be less sensitive to phosphorus supply than American beachgrass.

Management of this grass requires protection from traffic, prompt replanting of missing hills or breaks, and the introduction of other species able to take over and maintain stability in back-dune areas that are likely to become sand starved. Other grasses and legumes may be seeded in the intermediate zone for this purpose. Shrubs and trees are used farther inland,[4,21] but these do not tolerate foredune conditions.

#### 6. American Dunegrass

This plant (Figure 18) is the only foredune grass native to the northwest and is widely distributed throughout the region and occurs sporadically well southward along the Pacific.[9] It is primarily a foredune species, thriving under foredune conditions and requiring considerable annual sand deposition for healthy growth.[4] It builds foredunes. It is more difficult and expensive to propagate than European and American beachgrasses and consequently has not been planted to any extent. Its culture has received little attention until recently.[14] While American dunegrass appears not to be competitive with European beachgrass, it can be planted alone or in mixture with other native species to build foredunes. Due to its open spreading habit it produces low dunes with gentle slopes. This is often preferable to the high steep dunes formed by European beachgrass. A foredune in much of this region intercepts sand moving from the beach and prevents it from moving inland, rather than to bar storm tides. Here, there is no need for high, massive foredunes unless the space for sand storage is limited, then the type of dune from American dunegrass might be more appropriate. Use of this grass should be limited to the foredune zone because of high cost and plant requirements.

There is very little information available on the propagation and management of this species. The following suggestions are tentative and speculative.

### a. Planting Methods

American dunegrass should be planted in the same way as the beachgrasses. It should be set 30 cm or more deep in moist sand and the sand firmed around the plant to exclude air pockets. Barbour[14] found that adding peat moss to the planting hole improved survival. Dipping the plant bases in a clay slurry as is done with American beachgrass may be worthwhile. It is cheaper than peat moss and may serve the same purpose.

### b. Planting Date

This grass has been transplanted satisfactorily on the Oregon coast only when dormant (late November through February). It does not become dormant in winter at Pt. Reyes, Calif., which may partially explain poor survival there.[32] Temperature is also critical; planting is limited to temperatures below 13° C.

### c. Planting Stock

There is little information on plantings. Many stems grow from horizontal rhizomes or runners so more care should be taken with this species than with others. Because of the poor survival rate of this species, planting several stems per hill would be desired but may be too expensive. Closer spacing with one stem per hill makes better use of scarce planting stock.

### d. Spacing

Barbour[14] suggested planting at a density of 25 plants per square meter to attain comparability with natural stands. This would make planting too expensive for large-scale use. Other methods of thickening stands, such as the use of fertilizers, should be explored. Propagation techniques and more experience are needed to develop suitable planting patterns. However, the same general principles for spacing of other foredune grasses should apply with American dunegrass.

### e. Fertilization and Management

There are no published reports of fertilizer response of American dunegrass from the northwest. However, the European version of this plant (a subspecies or variety), lyme grass (*Elymus arenarius*), thickens markedly when fertilized.[7] Fertilizer should equally benefit new plantings of American dunegrass in this region. Its response to fresh sand strongly suggests that it would benefit from fertilization. A suggested application rate is 40 kg of nitrogen per hectare from a soluble source applied as soon as new growth starts.

Protection of new plantings from rabbits may be essential in some areas. Both American dunegrass and lyme grass are attractive to rabbits; small plantings would be particularly vulnerable. Other management requirements are the same as those described for the beachgrasses.

## 7. Other Plants

There are additional species occurring in each region that may have potential. These tend to be more restricted in range than the grasses.

Bermuda grass (*Cynodon dactylon*) is an introduced species which occurs frequently in the dune community in the South Atlantic region and less frequently in the North Atlantic. It can be used effectively for special purposes.

It is the best species for traffic-resistant turf on dunes. The turf-type hybrids, Tifway and Tifgreen, perform well throughout the South Atlantic when properly managed. Tufcote is used in the lower North Atlantic. These three hybrids are suggested for grassed walks, driveways, and parking areas. The hay-type hybrid, Coastal, roots well in dune sand and is quite salt-resistant.

Sprigging of plants is the usual method of establishment. Spacing is determined by rate of spread desired. For stabilization on bare foredune areas, 45 to 60 cm on centers is usually adequate. Closer spacing of 30 cm or less on centers should be used for turf. Sprigs for use in sand should be 15 to 20 cm long and set upright or at a slight angle with the tip, including some leaves or a joint, protruding above the surface. Sprigging may be done from early spring to summer under favorable moisture conditions. Early planting is usually best on foredunes or where supplemental water is not available. Where turf-type cover is required in a hurry and water is available, "instant" turf can be established by placing strips of sod over the area.

Bermuda grass has a much higher nutrient requirement than the typical dune grasses. Fertilization is the key to its establishment and maintenance. Suggested fertilization to develop plant cover on the foredune is 30 to 50 kg of nitrogen per hectare at 4-week intervals as soon as new growth begins in the spring through late summer. To develop traffic-resistant turf, 500 to 1000 kg of 10-10-10 fertilizer per hectare should be applied in early spring and followed by 50 to 75 kg of nitrogen per hectare at 4-week intervals through late summer.

The turf hybrids and Coastal variety will tolerate mowing but it should be infrequent and more top growth should be retained here than on inland sites to maintain trapping ability of sand.

Knot Grass or Seashore Paspalum (*Paspalum vaginatum*) is widely distributed along much of the South Atlantic coast, where it forms a turf on moist areas along road shoulders, beside ditches, and near dunes. It serves the same purpose on moister sites that Bermuda grass does in drier situations and it can be propagated and managed in the same way. It spreads naturally into suitable areas rather rapidly and does not usually require planting.

Seashore Elder (*Iva imbricata*) is the only broad-leaved plant with a potential for building and stabilizing foredunes in the South Atlantic. It is widely distributed throughout the region, although not generally plentiful. It occurs on backshores, foredunes, swales, back dunes, and in the upper fringe of the salt marsh where it is mixed with marsh elder (*I. frutescens*). It is highly tolerant to saltwater, salt spray, sandblast, and sand accretion. Occasionally, this plant alone builds foredunes (Figure 19) although it is usually mixed with one or more dune grasses. It spreads vegetatively and by seeds, and appears to be increasing.

Little was known about the ecological or propagation requirements of seashore elder until recently.[22] Not enough is known to predict future use, but the plant can be transplanted and contribute to attaining a more natural dune community. Community stability would be increased where it can be successfully introduced. It can grow throughout the pioneer and most of the intermediate zones.

Further research is needed for firm suggestions or recommendations. Three types of cuttings have been planted. Rooted stems taken from foredune plants have survived better than unrooted stems. Rooted cuttings in peat pots were more susceptible to wind

FIGURE 19.    Young dune initiated by seashore elder germinated in
wheel track.

erosion but those that survived grew faster than bare cuttings. Woody stems were better transplants than soft (new growth) stems.[23]

Seedlings only establish in areas of little sand movement and favorable moisture. Transplanting is successful on sand flats with a high water table. Transplanting to high, dry sites is not recommended. Seashore elder does not invade established foredunes but continues to grow with them after earlier establishment at lower elevations.

Pennywort (*Hydrocotyle* sp.) is a very effective sand-stabilizing broad-leaf plant. It is widely distributed throughout the region. It is tolerant of dune conditions, responsive to fertilization, and can be planted easily by sprigging. It is primarily a stabilizer rather than a builder because the round, fleshy leaves grow very close to the sand surface and provide only a few centimeters of trapping capacity at any one time. When only stabilization is needed, it can be sprigged in the same manner as Bermuda grass and fertilized like bitter panicum and sea oats.

St. Augustine Grass (*Stenotaphrum secundatum*) is a turf grass similar in habit and requirements to the turf-type Bermuda grasses. It is adapted to coastal conditions from about the southern quarter of the North Carolina coast southward. It can be substituted for turf-type Bermuda grass.

Spanish bayonet (*Yucca aloifolia*) grows well throughout the region and is useful as a windbreak. Sea grape (*Coccoloba uvifera*) is widely planted in central and south Florida as an ornamental windbreak.

Railroad vine (*Ipomoea pes-caprae*) is one of the more prominent plants along the southern 75% of the south Atlantic region. This is a robust vine that is capable of rapidly spreading over foredunes and back beach (Figure 20). It is not planted because it is less effective in trapping sand than the dune grasses. There are also other smaller *Ipomoea* sp.

There are several plants in the North Pacific that invade bare sand near the sea and build dunes.[24] The most prominent of these are yellow sand verbena (*Abrona latifolia*) and beach-bur (*Ambrosia chamissonis*). Sand verbena is salt tolerant but difficult to propagate. Beach-bur is easier to propagate but less tolerant of salt spray and inundation by seawater. Practical procedures for field planting are under study.[14]

Most of the plants that are capable of growing near the foredune can only be introduced after stabilization of the sand by pioneer species. These species are essential for sand stabilization whenever the sand supply declines. Seaside lupine (*Lupinus littoralis*) may be seeded into grass stands and does well as a soil improver. As it dies down in winter, it is not an effective stabilizer The beach peas (*Lathyrus japonicus* and *L.*

FIGURE 20.    Railroad vine in pioneer zone, Padre Island, Tex.

*littoralis*) grow well in this zone. Native grasses that can be seeded are seashore blue-grass (*Poa macrantha*), native creeping red fescue (*Festuca rubra*), sweet vernal (*Anthoxanthum odoratum*), and velvet grass (*Holcus lanatus*). Farther back from shore, shrubs and trees are effective as stabilizers.[21,24]

Three forms of seafig occur along the Pacific coast. They are the introduced Hottentot fig (*Carpobrotus edulis*), the native sea fig (*C. aequilaterus*) and hybrids between the two species. Hottentot fig is dark green or often red with stiff, pointed leaves 8 cm or more long, with white to pale-lavender flowers up to 8 cm wide. Native sea fig has fleshy, bluish-green soft leaves about 4 cm long, somewhat rounded at the tip, and pink flowers no more than 4 to 5 cm wide. The hybrids are usually intermediate in size, with pink flowers but with foliage resembling Hottentot fig.[25] Hottentot fig and the hybrids are very aggressive, covering dunes and many cliffs along this coast and excluding most other species, producing a rather monotonous and biologically uninteresting landscape. These types have usually been planted because of ease in planting and quick establishment.

Sea figs are effective sand stabilizers but not good dune builders. The growth above ground will not hold more than 10 to 15 cm of sand, usually less, at any one time and the plant cannot withstand much burial. It is only moderately tolerant to salt, and is susceptible to frost. Sea figs are the easiest plants to establish in the beach and dune zone. Cuttings 10 to 15 cm long set into moist sand about 45 to 60 cm apart quickly take root and provide cover. Successful stands of Hottentot fig have been established by broadcasting plants over the bare sand surface without covering during cool, rainy periods.[33]

Sea figs become nutrient-deficient on dune sands. The red color often exhibited by the exotic form is probably a response to low nitrogen levels. Occasional applications of nitrogen at the rate of 30 to 40 kg/ha are required to maintain vigor and good color. Applications should be adjusted to growth and appearance of the plants. Native sea fig may be readily transplanted to beach and dune areas. It is much slower growing and less vigorous than its exotic relatives and is not available commercially.

A number of other plants in the South Pacific region invade the foredune area and contribute in varying degrees to the building and stabilization of dunes. American dunegrass occurs sporadically but because of the difficulty of propagating this species in cooler climates, it should be used as a mixture with native plants.

Beach sagewort (*Artemisia pyenocphala*) is a pioneer species as far south as Monterey Bay. It can be readily transplanted, or seeded in less exposed areas. Divisions with roots and cuttings may be set in the moist sand.[25] Red sand verbena and beach-bur and, to a lesser extent, yellow sand verbena are common throughout the northern part of the region and are capable of invading bare sands near the beach. Yellow sand verbena is replaced entirely by red sand verbena south of Pt. Conception where the latter becomes the major foredune builder from there into Mexico.[32] Due to their growth habit, the sand verbenas never develop much capacity to trap sand but they do build dunes. These plants have tap roots; suitable field planting techniques for them have not yet been developed.[14] All of these species respond to moderate fertilization, particularly where fresh sand supply is meager or erratic.

Perennial veldt grass (*Ehrharta calycina*) was seeded at Vandenburg Air Force Base, Calif. on a large disturbed area immediately behind the dunes. A variety of temporary stabilization techniques was used. Straw mulch was anchored by discing, the surface was sprayed with bituminous materials, or the sand was irrigated by a sprinkler. The veldt grass was successful and the gradual reinvasion of native plants has occurred.[34] This grass appears to be very useful for initial sand stabilization.

There is considerable interest in the South Pacific region in the restoration of native sand dune plant communities as opposed to planting exotics to build and stabilize dunes. This approach has more promise in this region than in several others. Barrier dunes are not often required for storm surge protection as they are on low-lying coasts but rather are needed for the interception and storage of sand that would otherwise blow inland. Dune areas get very high recreational usage on this coast and native plants provide an aesthetically pleasing landscape. Protection from foot and vehicular traffic is required or restoration attempts are useless.

Cowan[25] described a procedure used to re-establish native species on active dunes at Asilomar State Beach, Pacific Grove, Calif. This method and its modifications are feasible where water for temporary irrigation is available and affordable. Briefly, the procedure involved temporarily stabilizing the dunes by hydromulching with a light seeding of annual and perennial ryegrasses, plus fertilizer. This was followed by periodic sprinkler irrigation to hold the sand and germinate the ryegrass, enabling it to grow into protective cover. Seeds of native plants could have been included in the hydromulch to accelerate the process but were not. Irrigation was terminated after 1 year and the ryegrass gradually died but protected the area for a second year. During the decline of the ryegrass, sea fig and beach sagewort were transplanted into the dying sod, and seeds of sand verbena, sand-bur, beach sagewort, seaside paintedcup (*Castilleja latifolia*), beach pea, sea rocket (*Cakile maritima*), and mock heather (*Haplopappus ericoides*) were broadcast over it. This approach developed a protective cover with a good variety of native species on the dunes at Asilomar State Beach within 4 or 5 years. It has promise for use in areas of limited size elsewhere. The principal problems are availability and cost of water, collection of transplants and seeds of native plants, and protection from foot and vehicular traffic.

There is a wide variety of native plants that occur in the pioneer zone around the Great Lakes; some are true pioneers, others are secondary or tertiary invaders.

Several species such as blue joint (*Calamagrostis canadensis*), prairie sandreed (*Calamovilfa longifolia*), tansy (*Tanacetom huronense*), European beachgrass (*Ammophilia arenaria*), false heather (*Hudsonia tomentosa*), and scouring rush (*Equisetum hyemale*) invade as pioneers and contribute to dune growth and stability. Sand cherry (*Prunus pumila*), creeping cedar (*Juniperus horizontalis*), forest grape (*Vitus riparia*), beach pea (*Lathyrus maritimus*), and willow (*Salix* sp.) are frequently secondary stabilizers.

Balsam popular (*Populus balsamifera*) is the most frequent tertiary species.

Most Great Lakes dunes can be converted to forest once they have been stabilized with herbaceous plants as trees can grow much closer to the shore because of the absence of salt. Fully established forests usually form the most durable, maintenance-free cover for Great Lakes dunes. Coniferous species have generally been the most successful on dune sands;[2,6] other trees and shrubs may be used to stabilize for special uses.[21]

### 8. Sand Fences

Establishment of vegetative cover on large areas of mobile sand is difficult, particularly under high-energy conditions. Success is not always possible on the first attempt and it may be necessary to supplement initial plantings with sand fences. These may be used to temporarily stabilize areas until plantings can become established. On low-lying sites, they are used to build a low ridge to raise planting sites above salt concentrations and storm tides.[8] Half fences (standard 1.2 m wooden slat fences cut in half) are usually better for use in conjunction with plantings.[26] Adequate guidelines on sand fence specifications, placement, function, efficiency, etc. have been developed.[20,26-29]

## IV. MANPOWER ANALYSIS

Obviously, manpower requirements will vary widely with the location, configuration, size, etc. of the planting site, as well as the experience of the operator and equipment. The following estimates were obtained from fairly large scale projects handled by experienced personnel.

Planting rate for American beachgrass is about 400 to 600 hills per man-hour by machine and 100 to 200 hills per man-hour by hand.[26]

Dahl[8] estimated harvesting, processing, and machine planting of bitter panicum at the rate of 230 hills per man-hour and 130 hills per man-hour for sea oats.

McLaughlin and Brown[4] reported a production rate of 136 to 156 hills per man-hour for hand planting European beachgrass. This is somewhat lower than east coast estimates for planting American beachgrass. Machine planting of European beachgrass should be possible at about the same rate as American beachgrass.

## V. COST

Planting stock together with labor will usually represent the major expense. This will vary widely with species, location, quantity required, and availability. Market demand for these materials is extremely erratic, depending to a large extent on storm events and the scheduling of major stabilization projects.

It is usually impractical to carry stocks in field nurseries for more than a year or two and it is not feasible or economical to hold over pot-grown seedlings from one planting season to the next. Consequently, there are few suppliers of these materials and much of their production is only on order. For plantings of any magnitude, it is essential from both the standpoint of cost and availability to plan ahead and place orders for planting stock well in advance.

American beachgrass is available on the Atlantic coast (1981 to 1982 season) at about $30 to $40/1000 sprigs.

The cost of European beachgrass planting stock is not available but since this is the easiest of the beachgrasses to propagate, planting stock should be about the same or cheaper than American beachgrass.

Pot-grown sea oats seedlings are available, produced to order. These are considerably more expensive than the field-grown *Ammophilia* stock.

Fertilization costs (materials, rates, and application methods) vary considerably from site to site and region to region. In 1981, enough nitrogen (40 kg) to fertilize a hectare could be purchased as ammonium nitrate for $30 to $40. Application can easily cost as much or more than the materials.

Sand fence, where needed, will cost around $2.00/m for the standard 1.2 m wooden slat fence; material for posts and braces probably less than $1.00/m of fence; and one man can install about 9 to 14 m/hr.

## VI. FACTORS CONTROLLING SUCCESS OR FAILURE

Coastal dunes usually occur in high energy, exposed situations and attempts at creating dunes and restoring dune vegetation often fail. These failures are frequently due to uncontrollable factors such as blow-outs, excessive sand burial, or washouts and about the only cure for these is to keep trying. Sites that are impossible one year may often be readily planted the next season. There are sites that have been very successfully restored on the fourth, fifth, or even sixth attempt.

There are also controllable factors that account for many failures. The most frequent are old, stunted, weak planting stock, improper handling and storage of plants, poor transplanting procedures — usually shallow planting and inadequate firming of sand around the transplant base. In the absence of fresh sand, inadequate nutrient levels are a frequent cause of failure.

Close and constant attention to the details of plant quality and planting procedures is the best insurance against failures.

Location of a barrier dune can have a major influence on its durability and utility. Well-vegetated dunes are effective against storm tides and are capable of withstanding moderate degrees of overtopping, but they are highly vulnerable to undermining through beach recession and persistent wave attack. In the placement of a new barrier dune, an allowance should be made for the normal shoreline fluctuations characteristic of the site. Serious problems of dune maintenance may often be avoided or minimized by locating the foredune back from mean high water (MHW) far enough to allow for a reasonable amount of seasonal fluctuations. Dutch workers[30] suggest that the minimum distance between the toe of the dune and MHW should be 200 m.

It is also important to consider the nature of dune growth in locating a barrier dune. Fully vegetated dunes expand only toward the sand source, usually the beach, and a relatively narrow strip of vegetation will, in most cases, stop all wind-transported sand. This means that, where possible, allowance should be made for seaward expansion of the dune with time. Also, if two dunes are desired, the first must be developed landward and enough space left between it and the sea for the second or frontal dune (Figure 14).

On many low-lying coasts the crest of the storm berm is the highest point in the beach-dune area with the surface sloping back from it. This places the base of a new barrier dune below the elevation of the storm berm, making it more susceptible to overtopping during the early stages. It may also encourage ponding of water coming over the storm berm, resulting in water pressure, salt buildup, and destruction of vegetation along the toe of the dune. Where this problem exists, dune location must always represent a compromise.

## VII. RESEARCH NEEDS

The three major problems on which further research and development is needed are (1) reduction in planting costs particularly with certain species such as sea oats, (2) development of propagation and planting requirements and procedures for less frequently planted species such as American dunegrass and some of the tap-rooted species like sand verbena, and (3) selection and evaluation of ecotypes best suited for different areas. For example, there is a wide genetic variation within bitter panicum.[31] There are very likely different ecotypes that would be much more useful than others in different parts of the region.

A special problem is the need for a companion grass to grow with American beachgrass in the north Atlantic region where it so severely decimated by scale.

# REFERENCES

1. **Westgate, J. M.,** Reclamation of Cape Cod sand dunes, Bull. No. 65, Bureau of Plant Industry, U.S. Department of Agriculture, Washington, D.C., 1904.
2. **Lehotsky, K.,** Sand dune fixation in Michigan, *J. For.,* 39(12), 998, 1941.
3. **Meyer, A. L. and Chester, A. L.,** The stabilization of Clatsop Plains, Oregon, *Shore Beach,* 45(4), 34, 1977.
4. **McLaughlin, W. T. and Brown, R. L.,** Controlling coastal sand dunes in the Pacific Northwest, Circ. No. 660, Department of Agriculture, Washington, D.C., September 1942.
5. **Woodhouse, W. W., Jr. and Hanes, R. E.,** Dune stabilization with vegetation on the Outer Banks of North Carolina, TM 22, U.S. Army, Corps of Engineers, Coastal Eng. Res. Cent., Washington, D.C., August 1967.
6. **Lehotsky, K.,** Sand Dune Fixation in Michigan — Thirty Years Later, *J. For.,* 70(3), 155, 1972.
7. **Adriani, M. J. and Terwindt, J. H. J.,** *Sand Stabilization and Dune Building,* Rijkwaterstaat Communications, No. 19, The Hague, The Netherlands, 1974, 68.
8. **Dahl, B. E., Bruce, A. F., Lohse, A., and Appen, S. G.,** Construction and stabilization of coastal foredunes with vegetation: Padre Island, Texas, MP 9-75, U.S. Army, Corps of Engineers, Coastal Eng. Res. Cent., Fort Belvoir, Va., September 1975.
9. **Barbour, M. E., DeJong, T. M., and Johnson, A. F.,** Synecology of beach vegetation along the Pacific Coast of the United States of America: a first approximation, *J. Biogeogr.,* 3(3), 55, 1976.
10. **Woodhouse, W. W., Jr., Seneca, E. D., and Broome, S. W.,** Ten years of development of man-initiated coastal barrier dunes in North Carolina, Bull. No. 453, North Carolina Agric. Exp. Stn., Raleigh, N.C., December 1976.
11. **Campbell, W. V. and Fuzy, E. A.,** Survey of scale insect *Eriococcus carolinae* Williams, *Shore Beach,* 40(1), 18, 1972.
12. **Lucas, L. T., Warren, T. R., Woodhouse, W. W., Jr., and Seneca, E. D.,** Marasmius blight, a new disease of American beachgrass, *Plant Dis. Rep.,* 55(7), 582, 1971.
13. **Seneca, E. D., Woodhouse, W. W., Jr., and Broome, S. W.,** Dune stabilization with *Panicum amarum* along the North Carolina coast, MR 76-3, U.S. Army, Corps of Engineers, Coastal Eng. Res. Cent., Fort Belvoir, Va., February 1976.
14. **Barbour, M. E.,** Management of dune and beach vegetation, Sea Grant Project R/CZ-22, Annu. Rep., University of California, Davis, 1976.
15. **Seneca, E. D.,** Germination and seedling response of Atlantic and Gulf coasts populations of *Uniola paniculata, Am. J. Bot.,* 59, 290, 1972.
16. **Brown, R. L. and Hafenrichter, A. L.,** Factors influencing the production and use of beachgrass and dunegrass clones, *Agron. J.,* 40(6), 512, 1948; 40(7), 603, 1948; 40(8), 677, 1948.
17. **Berenyi, N. M.,** Soil Productivity Factors on the Outer Banks of North Carolina, Ph.D. dissertation, North Carolina State University, Raleigh, unpublished, 1966.
18. **Jagschitz, J. A. and Bell, R. S.,** Restoration and retention of coastal dunes with fences and vegetation, Bull. 382, Rhode Island Agric. Exp. Stn., Kingston, R. I., 1966a.
19. **Zak, J.M.,** Controlling drifting sand dunes on Cape Cod, Bull. No. 563, Massachusetts Agric. Exp. Stn., Amherst, Mass., 1967.
20. **Savage, R. P. and Woodhouse, W. W., Jr.,** Creation and stabilization of coastal barrier dunes, in *Proc. 11th Conf. Coastal Eng.,* London, Vol. 1, American Society of Civil Engineers, 1968, 671.
21. **Brown, R. L. and Hafenrichter, A. L.,** Stabilizing sand dunes on the Pacific Coast with woody plants, Misc. Publ. 892, U.S. Department of Agriculture, Washington, D.C., 1962.
22. **Colosi, J. C. and McCormick, J. F.,** Population structure of *Iva imbricata, Bull. Torrey Bot. Club,* 105(3), 175, 1978.
23. **Colosi, J. C., Seneca, E. D., and Woodhouse, W. W., Jr.,** Dune stabilization with *Iva imbricata,* Botany Department, North Carolina State University, Raleigh, in preparation.
24. **Wiedemann, A. M., Dennis, L. R. J., and Smith, F. H.,** Plants of the Oregon coastal dunes, Department of Botany, Oregon State University, Corvallis, 1974.
25. **Cowan, B.,** Protecting and restoring native dune plants, *Fremontia,* 2, 3, 1975.
26. **Woodhouse, W. W., Jr.,** Dune building and stabilization with vegetation, SR-3, U.S. Army, Corps of Engineers, Coastal Eng. Res. Cent., Fort Belvoir, Va., September 1978.
27. **Bagnold, R. A.,** *The Physics of Wind Blown Sand and Desert Dunes,* Methuen and Company, London, 1941.
28. **Manohar, M. and Bruun, P.,** Mechanics of dune growth by sand fences, *Dock Harbour Auth.,* 51, 243, 1970.
29. **Phillips, C. J.,** Review of selected literature on sand stabilization, Department of Engineering, University of Aberdeen, Scotland, 1975.
30. **Blumenthal, K. P.,** The construction of a draft sand dyke on the Island Rottmerplatt, in *Proc. 9th Conf. Coastal Eng.,* American Society of Civil Engineers, 1964, 346.

31. **Palmer, P. G.**, A biosystematic study of the *Panicum amarum p. amaralum* Complex (gramineae), *Brittonia,* 27(2), 142, 1975.
32. **Johnson, A. L.**, personal communication, 1977.
33. **Crisp, E. T.**, personal communication, 1976.
34. **Moore, C. L.**, personal communication, 1976.

## GENERAL REFERENCES

**Craig, R. M.**, Coastal dune vegetation, Proc. Fla. State Hortic. Soc., 1974.
**Davis, J. H., Jr.**, Stabilization of beaches and dunes by vegetation in Florida, Rep. No. 7, Florida Sea Grant Program, Gainesville, September 1975.
**Graetz, K. E.**, Seacoast plants of the Carolinas, U.S. Department of Agriculture, Soil Conservation Service, Raleigh, N.C., February 1973.
**Menniger, E. A.**, *Seaside Plants of the World,* Heatbridge Press, New York, 1964.
**Schory, E. A.**, Salt tolerant, cold hardy, drought resistant trees and shrubs for North and Central Florida, Division of Forestry, Fort Meyers, Fla., 1970.

Chapter 2

## ATLANTIC COASTAL MARSHES

### W. W. Woodhouse, Jr. and Paul L. Knutson

## TABLE OF CONTENTS

# I. INTRODUCTION

## A. Natural Plant Communities

A coastal marsh is a herbaceous plant community found on the part of the shoreline which is periodically flooded by salt or brackish water. A number of species in the grass family (Poaceae), sedge family (Cyperacea), and rush family (Juncaceae) commonly form coastal marshes. Coastal marshes occur naturally in the intertidal zone of moderate to low-energy shorelines along tidal rivers and in bays, estuaries, and sounds. These marshes may be narrow fringes along steep shorelines but can extend over wide areas in shallow, gently sloping bays and estuaries.

Vegetation of east coast marshes is remarkably uniform. The intertidal zone from New England to Florida is dominated by a single species, smooth cordgrass (*Spartina alterniflora*) (Figure 1). Two grasses, saltmeadow cordgrass (*S. patens*) and saltgrass (*Distichlis spicata*), usually dominate the zone immediately above high tide along these coasts with two rushes on slightly higher sites — black-grass (*Juncus gerardi*) north of the Virginia Capes and black needle rush (*J. roemerianus*) southward.

East coast marshes divide into three general types: New England, mid-Atlantic, and south Atlantic. Typical New England marshes occur on fibrous or silty peat.

The intertidal zone of pure stands of smooth cordgrass is usually relatively narrow with a well-developed upper zone of saltmeadow cordgrass mixed with salt grass (Figure 2). Saltmeadow cordgrass often occupies a larger area than smooth cordgrass. Pure stands of black-grass in the higher parts of the zone often form a fringe at the edges of the uplands.

Marshes in the mid-Atlantic region undergo subtle changes from the New England type on Long Island to the South Atlantic type at the Virginia Capes. There are relatively limited areas of smooth cordgrass with the greatest area covered by saltmeadow cordgrass. Localized high salinity patches are dominated by pickleweed (*Salicornia* sp.). Big cordgrass (*Spartina cynosuroides*) and several rushes (*Scirpus* sp.) occur along creeks and tidal stream mouths where the freshwater influence is greater. Black needle rush increases in importance near the mouth of the Chesapeake Bay. The tall form of smooth cordgrass appears along creek banks.

South of the Chesapeake Bay, the South Atlantic marshes typically form behind barrier beaches and in estuaries where rivers deposit heavy silt burdens. Smooth cordgrass occupies vast areas of mostly soft sediments between mean sea level and mean high water. Large areas of high marsh, primarily black needle rush, occur where astronomical tides are restricted and wind setup predominates; e.g., in Pamlico Sound, N.C. South of Daytona Beach, Fla. the typical South Atlantic marshes are largely replaced by mangrove trees that form the tropical and subtropical equivalent of salt marshes.

### B. Historic Losses

Until recently, these wetland resources have been steadily shrinking. They were generally considered useless and viewed as prime areas for agricultural, commercial, and recreational uses and for waste disposal. Nationwide coastal marshes are being lost at a rate of about 1/2%/year. Since 1954, 1/2 million ha have been lost.[1] The 1 million ha of marsh that remain on the Atlantic coast constitute the most extensive salt marsh system in the world. Wetland losses vary greatly from state to state. By the 1950s Connecticut had lost nearly one half of its coastal wetlands. Since then, nearly half of that which remained has disappeared. In Georgia, where marshlands are abundant and coastal development has been a relatively recent phenomenon, the loss may be only 5%.[2] However, a simple comparison of historic and current marsh abundance may understate the total damage which has been done to these environments. Many remaining wetlands have been severely modified. It has been estimated that 90% of the tidewater marshlands between Maine and Virginia have been modified by ditching and other human activities.[3]

### C. Values

Destruction of coastal wetlands has lessened as the value of these areas as nursery grounds and sources of primary production (energy) for a high proportion of sports and commercial fishery species has become widely recognized.[4-8] Most species of sport and commercial fishes spend at least a portion of their life cycle in estuaries where they depend on marshes. Salt marshes may produce as much as 30,000 kg/ha of organic matter per year. This production is distributed to coastal waters by tides and currents. Marsh vegetation, phytoplankton, and seagrasses are the crops of the sea. They are the vehicles by which the energy of the sun is ultimately converted to animal protein: shrimp, oysters, fish, shorebirds, etc. For example, a recent study found that shrimp catches, worldwide, are directly related to the area of marsh in the shrimp nursery grounds.[9] The destruction of North Atlantic coastal marshes is at least in part responsible for the 80% decline in commercial fishery landing of estuarine-dependent species since 1920 in Connecticut, New York, and New Jersey[10]

### D. Marsh Restoration

Interest has developed in marsh restoration, in the building of new marshes to replace those that have been lost, and in the use of marsh plants to stabilize and protect eroding shorelines. Studies on marsh building were initiated in North Carolina in 1969 under the sponsorship of the Coastal Engineering Research Center[11-13] and later expanded to the Chesapeake Bay[14] and the Gulf coast.[15,16] The U.S. Army Engineer Waterways Experiment Station, Vicksburg, Miss. has made marsh-building tests on dredged material at several widely distributed locations.[17] These studies, and earlier plantings along tidal river shores in Virginia[18] have demonstrated the feasibility of establishing new coastal marsh under a variety of situations. Knutson[19] located and evaluated more than 70 coastal marsh plantings on the Atlantic coast.

Though the past decade of research has greatly advanced our understanding of how

FIGURE 1.    Smooth cordgrass marsh, Virginia.

FIGURE 2.    Saltmeadow cordgrass in foreground; smooth cordgrass in background — Maine.

coastal marshes can be created or restored, this is still a very new field. The guidelines presented in this chapter for planting these environments are based upon both experience and current literature. Existing information will require considerable extrapolation and many of the resulting recommendations will be speculative. Woodhouse[20] summarizes planting guidelines for coastal marshes. His report is the primary source of information for this chapter, though additional information was obtained from Knutson and Woodhouse.[21]

## II. PLANT MATERIALS

A wide variety of plants can grow under marshy conditions where the water is fresh or only mildly brackish. This chapter emphasizes the plants that, at present, appear useful in coastal marsh creation. The small number of these plants is due to the saline conditions prevailing in most coastal marshes, the rigorous conditions during establishment, the difficulties encountered in propagating some species, the lack of information

FIGURE 3.    Smooth cordgrass.

available on others, and the secondary role a number of the plants play in stabilization and productivity. A great deal is known about where and under what conditions many marsh plants grow. Interest in planting them is of recent origin and planting requirements are known for only a few. More species are likely to be found useful in the future.

In this section, six coastal marsh species will be discussed individually: smooth cordgrass, saltmeadow cordgrass, black needle rush, saltgrass, big cordgrass, and common reed. Additional species also found in Atlantic marshes will be discussed briefly at the end of this section. The general topic of marsh plant propagation has been reviewed by Kadlec and Wentz.[22]

## A. Smooth Cordgrass (*Spartina alterniflora*)
### 1. Plant Description
This is the dominant flowering plant in the regularly flooded intertidal zone along the Atlantic coast from Newfoundland to about Central Florida (Figure 3). These marshes are essentially pure stands of smooth cordgrass. This grass is well-adapted to sea strength salinity (35⁰/₀₀), excreting salt through salt glands in its leaves. The plant is also well-adapted to the anaerobic substrates characteristic of most salt marshes. Its oxygen transport system consists of hollow, air-filled tissue, extending from openings in the leaves to the roots and rhizomes.[23,24] Thus, oxygen reaches the below-ground

tissues in anaerobic substrates. This grass can grow in a wide range of substrates from coarse sands to silty clays. Although dominant in regularly flooded, saline habitats, it is not restricted to these areas. It usually attains maximum growth under lower salinities (10 to 20°/oo). The grass will grow and reproduce normally under freshwater conditions but is subject to increasing competition from other species as salinity declines.[11,12]

Three distinct height forms (short, medium, and tall) covering a range of about 0.5 to 3 m, have been widely recognized. Within a natural marsh, the tall form occurs along tidal creeks and drainage channels, and the short form on flat or very gently sloping areas away from the channels. The medium height form, when present, usually occupies a band between the tall and short forms. It is uncertain whether the differences in growth habit and productivity are due to genetic factors or the result of local environmental conditions. Earlier workers[25,26] suggested that the stunted form is a genetic variety. More recent greenhouse,[27] biochemical,[28] and field transplant studies[13] indicate that these differences are largely, if not altogether, environmental.

There are, however, distinct geographic populations of smooth cordgrass. Seneca[29] grew plants from seeds collected from Plum Island, Mass. to Port Aransas, Tex. under controlled conditions and in the field at Snow's Cut, N.C. and found that there were at least four groups differing in time of flowering, growth, reaction to photoperiod, culm (stem), and leaf color. Flowering progressed from north to south, growth was adapted to a progressively longer growing season north to south, and basal culm diameter and leaf width increased from north to south.

Vegetative reproduction by extensive below-ground, hollow stems (rhizomes) is the primary method of spreading in established stands. Although seed production is usually limited in old dense stands, it may be substantial in newly established stands and along margins such as the borders of tidal creeks. Seeds are important in spreading the plant into new areas and often contribute to thickening of open or patchy stands.

Smooth cordgrass has probably received more study and can be planted with better chance of success than any other coastal marsh species, native to the U.S. It is relatively easy to propagate and quick to establish and spread. This grass tolerates inundation better than any other salt marsh species on the Atlantic coast. Consequently, it is valuable in protecting the lower slope of spoil disposal areas and eroding shorelines.

### 2. Cultural Techniques

Smooth cordgrass is propagated by seeds, by vegetative materials harvested from natural stands, or by vegetative materials produced in nurseries.

### a. Seeding

Seed production is confined largely to new, open stands and along margins; e.g., along tidal creeks. The most vigorous stands usually produce the best seeds but variability is high. Planted areas usually yield heavy seed crops for several years following establishment. Seed heads are frequently damaged by ergot infestation and by flower beetles (family Mordellidae).[30] Flowering time and seed maturity progress from north to south, at least within geographic populations such as along the Virginia-Carolinas coast. For example, there is a spread of about 2 weeks, north to south, in seed maturity along the North Carolina coast with considerable variability within individual stands. Seeds are ready for harvest as early as September in northern latitudes and as late as November in the south Atlantic marshes but maturity varies from year to year.

Harvest (cut and collect seed heads) is done by wading or from boats. This must be done shortly after maturity when seeds can be readily dislodged from the heads by rubbing, because they shatter readily soon after. Heads should be stored moist, but not submerged, at 2 to 3°C, for 2 to 3 weeks to allow "after ripening". They may then be threshed to reduce storage space and to facilitate handling, and stored in water of 20 to 25⁰/₀₀ salinity at 2 to 3°C until planting time.[12] Submerged storage is required because dry seeds lose viability rapidly,[27] and saline water is preferable, at least in some instances.[12] Low temperatures during storage are essential to retard germination as sprouting of ripe seeds is rapid under high temperatures following after-ripening. Even under the best storage conditions, large numbers of seed will sprout by the following March or April. These sprouted seeds are still usable for planting but are much more susceptible to damage from handling than unsprouted seeds. Freezing, either wet or dry, is not a satisfactory method of storage.[12] Viability of stored seed is not retained longer than 1 year. Consequently, seed must be harvested each year in September to November for planting the following year.

Smooth cordgrass invades new sites primarily by seeds; stands can be established by direct seeding on the more protected sites.[11,12] When feasible, this will usually be the most economical method. However, vegetative transplants are much more tolerant of waves and currents and should be used on most sites.

### b. Field Harvesting

Field collected plants are satisfactory and often adequate for small-scale plantings. These should come from uncrowded stands. This usually means stands of recent origin. Plants are obtained by loosening individual clumps with a shovel, small back hoe, or plow, and lifting and separating into individual transplants. Choice transplants consist of large, single stems (culms) with small shoots and short pieces of rhizomes left attached or discarded. Digging and processing of planting stock from old, dense marshes is difficult and usually yields small, poor quality plants. Where planting stock must be obtained from such stands, it is usually preferable to resort to plugs or cores as these small, single stems are not satisfactory as transplants.[12] Heavy harvest of single-culm plants initially appears to be devastating to the stand. However, the effect is very short-lived, particularly in open, vigorous stands on sandy substrates; remaining rhizomes and shoots soon repopulate the area, usually in the same growing season. It is difficult to harvest such sites close enough to prevent overcrowding and the reduction in suitability of the planting stock in succeeding years.

Due to the rapid recovery of vigorous new stands, the harvesting of planting stock from year-old plantings for marsh building is often feasible. Such stands yield excellent quality transplants at low cost with only a slight delay in the process of marsh development.

Plugs are obtained by excavating cubes or cylinders containing crowns, stems, roots, rhizomes, and soil from a healthy stand of cordgrass growing on a silty or clay substrate. Diameter of plugs must usually be 12 cm or more in order to include one or more intact plants. This form of planting stock is considerably more laborious to harvest and transplant but is the only feasible type where plants must be obtained from old crowded stands on heavy-textured substrates.[31-33]

### c. Nursery Materials

Field nurseries are relatively easy and economical to establish if suitable sites are available. An ideal site is a bare, smooth, intertidal slope of sandy material along a relatively protected shore. The initial stand may be established by seeding or transplanting single stems. Seeding rate should be low and transplants spaced at least 1.4 m apart. Row planting would facilitate mechanization of harvest operations. Diked pond sites, constructed of dredged materials, make excellent nursery sites if provided with a suitable water supply. Field nurseries are planted in the spring, and planting stock is available for harvest the following late winter, spring, and summer. Although this method has not been widely used, it has potential in many areas under periodic

FIGURE 4.  Nursery seedlings (Environmental Concern, Inc., St. Michaels, Md.)

dredging operation. Dredged material can be deposited to form an intertidal slope, planted with seeds or transplants in the spring, harvested for planting stock the following winter, spring, and summer, and remain thereafter as an addition to the marshlands of the area. Alternatively, it could be reactivated as a nursery in later years by covering the surface with a thin layer of sandy dredged material. This has not been done on an organized basis, but the effect has been observed where dredged material was deposited on established marsh. With proper depth coverage, about 8 to 12 cm, vigorous new culms that make good transplants emerge.

Seedlings in peat pots are produced in a greenhouse or out-of-doors during mild weather (Figure 4). Seeds are planted directly in sandfilled peat pots or germinated in flats and transplanted to pots later. Garbisch, Woller, and McCallum[14] grew four to six seedlings in each 5- to 10-cm pot. They fertilized with Hoagland's nutrient solution[34] (commercial fertilizers are also suitable) and adjusted salinity with sodium chloride. They state that salinity conditioning is necessary if seedlings are to be transplanted in salinities above 15°/$_{oo}$. Garbisch[35] recommends planting potted seedlings 15 weeks old in sheltered areas and 5- to 7-month old seedlings on moderately exposed to exposed sites. Pot-grown transplants suffer less root damage than plants dug from the field and are able to resume growth quicker following transplanting. This extends the planting season by giving the plants a longer period in which to become established. However, peat-pot seedlings are bulky and much heavier than field or nursery transplants, are more difficult to transport, and have more exacting transplanting requirements.

## B. Saltmeadow Cordgrass (*Spartina patens*)
### 1. Plant Description
This is a fine-leaved grass, 15 to 80 cm in height, that occurs extensively in the irregularly flooded high marsh zone all along the Atlantic coast. In the absence of black needle rush, it replaces smooth cordgrass at about mean high water (MHW) and forms dense mats from MHW to the high spring or storm tide line. This grass often forms a narrow band along the marsh edge, but on gently sloping topography, it may cover a wide expanse and be mixed with saltgrass, patches of needle rush, and other high marsh species. Saltmeadow cordgrass forms the extensive saltmeadows of New England that were formerly mown for hay. This grass also occurs at higher elevations on sandflats and low dunes where growth is sparse.

Saltmeadow cordgrass can withstand extended periods of both flooding and drought, and often occurs where surface drainage is poor, causing ponding of rainwater during wet periods. It cannot tolerate the daily flooding of the intertidal zone. Productivity can be high, but this species' contribution to the detrital food chain is much less direct than that of smooth cordgrass.

Saltmeadow cordgrass is a valuable stabilizer for the zone between the smooth cordgrass and the high spring or storm tide line or the zone of adaptation of the upland grasses such as tall fescue (*Festuca arundinacea*), bermuda (*Cynodon dactylon*), and St. Augustine (*Stenotaphrum secundatum*). It is relatively easy to multiply and transplant.

### 2. Cultural Techniques

This grass is propagated vegetatively and by seeds. It seeds profusely, often invades low, moist sites by seed, and can be directly seeded on protected sites. However, transplanting is preferable on more exposed or steeply sloping areas that may be subject to erosion.

### a. Seeds

Saltmeadow cordgrass is a fairly consistent producer. It grows on irregularly flooded and unflooded sites, and the seeds do not require moist storage.[16] Large-scale harvesting and processing of this species could be handled with the same equipment and in a similar manner as many of the cultivated grasses. Small quantities are harvested by hand as with smooth cordgrass. Seed should be stored dry. Storage at low temperature is probably best, although there is no clear-cut evidence to support this.

### b. Field Harvesting

This plant is plentiful in high marshes and on low sandflats along the Atlantic and Gulf coasts, but it is difficult to obtain good planting stock from the wild. Stands on moist sites soon become so dense that harvesting is difficult and the crowded plants do not make vigorous planting stock. Plants growing on dry, infertile sites lose vigor and survive poorly when transplanted. The best transplants are the large culms from rapidly growing, uncrowded young stands; however, obtaining significant quantities of this kind of transplant in most areas will require the establishment of a nursery.

### c. Nursery Materials

Saltmeadow cordgrass can be grown as readily inland as on the coast. Plant on a weed-free, sandy soil with a good moisture-holding capacity. The seedbed should be well pulverized and, if needed, fumigated with methyl bromide to kill weed seeds. Seed may be used for nursery establishment, but transplants are usually more practical. Nursery plantings should be made in late winter or spring. Use one to three-culm transplants from young, vigorous stands, set 10 to 15 cm deep in moist soil, 45 to 60 cm apart in rows. Space rows to allow cultivation, usually 75 to 110 cm apart; fertilize at planting; topdress with nitrogen later if need is indicated by growth and appearance.

It is usually best to harvest nursery-grown stock after one growing season to avoid the development of overcrowded, less desirable plants. Harvest by loosening individual clumps with a shovel, a tree digger, or a similar tool and then lift. Saltmeadow cordgrass culms are small even under the best growing conditions; clumps should be divided into four to eight-culm plants for transplanting. Plants may be placed upright in tubs, baskets, or crates for handling and transport, or bundled in the same way as tree seedlings. Care must be taken to avoid drying and heating. Plants may be heeled-in in moist sand for temporary storage.

Saltmeadow cordgrass may be grown in peat pots in the same general way as de-

FIGURE 5.    Dark clumps in photo center are six single-stem plantings of black needle rush after five growing seasons, North Carolina.

scribed for smooth cordgrass. In view of the ease of field propagation of this species,
there appears to be less justification for this more costly method. However, where salt
buildup is likely to interfere with initial establishment, the more intact root systems of
the peat-pot seedlings should have an advantage. Salt buildup is likely in parts of the
saltmeadow cordgrass zone when inundation by spring or storm tides is followed by
periods of low rainfall and warm temperatures. Established plants can tolerate this,
but fresh transplants may be severely damaged.

## C. Black Needle Rush (*Juncus roemerianus*)

### 1. Plant Description

Black needle rush has stems and leaves that are round in cross-section, rigid, with
sharp-pointed tips capable of penetrating the skin. Dense stands have a brown to gray-
black appearance with little change in color throughout the year. Height ranges from
0.5 to 1.5m. This plant occurs extensively along the Atlantic coast south of New England as high marsh just above MHW, flooded only by wind-driven tides. It also grows
in mixture wth smooth cordgrass and saltmeadow cordgrass and in extensive stands
near the edge of the uplands where there is regularly seepage of freshwater. Productivity of black needle rush can be fairly high; however, there is little transfer of biomass
from these marshes to the estuaries. Old growth tends to remain standing for 1 year
or more. Much of the production goes into peat formation rather than the estuarine
food chain. Growth and spread of needle rush is slow and protracted. A single stem
planting may require 5 years to spread a distance of 1 m. Once established, however,
it will thwart most attempts to displace it (Figure 5).

### 2. Cultural Techniques

This plant is propagated vegetatively and by seeds. Transplant success has been erratic. Plants from young, uncrowded stands are definitely preferable to older plants.
Seeds may germinate as soon as shed. They require light and constant wetness, and
germinate best in fresh water; prolonged exposure to salinities above 1% are detrimental. Black needle rush seeds are more difficult to harvest than seeds of the cordgrasses.
Seedlings have been produced in peat pots[14] which is probably the most reliable
method. However, in light of the difficulties encountered in direct establishment of
this species and the propensity it has for invading stands of other marsh plants after

stabilization, direct planting of black needle rush is seldom justified. Usually, it is much easier to stabilize the area with smooth cordgrass or saltmeadow cordgrass and allow black needle rush to invade naturally where it is best adapted. If large grass plantings are isolated from natural stands, it may be advisable to include 1 to 5% black needle rush in tbe initial planting to ensure the presence of a seed supply.

### D. Saltgrass (*Distichlis spicata*)

This grass is widely distributed in high marshes along the Atlantic Coast. It is a low-growing grass (0.1 to 0.4 m high), with a pale or whitish-green cast. It is rarely dominant except in small poorly drained, more saline patches, and usually occurs mixed with saltmeadow cordgrass or rushes. Saltgrass is an effective stabilizer and is more salt tolerant than most other high marsh species.[37] It is more difficult to establish than the cordgrasses but readily volunteers into areas first stabilized with other marsh plants. This and the fact that it is rarely dominant suggest that saltgrass should not normally be direct-planted but rather allowed to volunteer into those parts of a planting to which it is best adapted.

This plant spreads vegetatively and by seeds. Transplanting success using sprigs has been poor. Survival has been low and initial growth slow. A recent report[38] stated that this grass survived well, spread rapidly the first 2 years after transplanting, and was then gradually replaced by taller species. Results have been obtained with peat pot-grown seedlings.[14] Because saltgrass readily invades established stands of other marsh species, artificial propagation of this plant is seldom worthwhile.

### E. Big Cordgrass (*Spartina cynosuroides*)

This grass is taller (1 to 3 m), with larger leaves, stems, and rhizomes than smooth cordgrass. It grows in salty or brackish areas, above about mean high water. Big cordgrass forms a dense root-rhizome mat and is a good stabilizer but not as effective as saltmeadow cordgrass; it dies back during cold weather. The grass covers large areas of high marsh along brackish shores such as in Currituck Sound, N.C. However, because it does not extend much below mean high water, it cannot protect the lower slope and is often undermined by waves in that zone. Propagation of this plant is similar to smooth cordgrass but with much more difficulty. Consequently, like black needle rush and black grass, indirect establishment of big cordgrass by planting other easier handled species, such as smooth and saltmeadow cordgrass, seems to be more practical.

This plant spreads vegetatively and by seeds. Plants have been grown from seeds in peat pots.[35] Planting has also been done using transplants collected from the wild. However, this plant does not transplant as readily as smooth and saltmeadow cordgrasses and it is essential to use material from young, uncrowded stands for transplanting. These stands are usually scarce and difficult to find. In view of the specific elevation and flooding requirements of big cordgrass, suitable nursery sites will probably be difficult to find and develop. Consequently, planting stock production by seeding in peat pots is probably the most feasible method.

### F. Common Reed (*Phragmites australis*)

This large coarse grass is 1.5 to 4 m tall and widely distributed in brackish to fresh-water areas where it grows above about MHW (Figure 6). Common reed seeds profusely, spreads vegetatively by rhizomes and stolons, is easy to transplant, and is a good stabilizer. Where adapted, it grows and spreads vigorously, often excluding other species. It is not generally favored as wildlife food or cover although it is eaten by muskrats. Since it can become a nuisance by crowding out more desirable plants, it should be introduced into new areas with extreme caution. It loses some stability when it dies back during the winter.

FIGURE 6.    Common reed planting, North Carolina. (Courtesy of U.S. Army, Corps of Engineers, Coastal Eng. Res. Cent., Fort Belvoir, Va.)

To plant, simply divide stems, rhizomes, or stolons into segments containing at least one node. Each node will produce a new plant if planted in a proper environment. It is unlikely that any situation will arise that would justify nursery cultivation.

## G. Other Plants

Sea oxeye (*borricha frutescens*), marsh elder (*Iva frutescens*), goldenrod (*Soldago* sp.), sea myrtle (*Baccharis halimifolia*), and mallow (*Hibiscus* sp.) are secondary plants of the high marsh. Most of these could be planted but will normally invade planted areas when and where conditions are favorable. Pickleweed and sea lavender (*Limonium* sp.) are common in the higher parts of the low marsh with pickleweed often in areas of salt concentration.

Saw grass (*Cladium jamaiecense*) occupies mildly brackish areas. Cattail (*Typha* sp.) is a common freshwater marsh species and has been planted for marsh building.[39]

Arrow-arum (*Peltandra* sp.) and pickerelweed (*Pontederia cordata*) have been planted with some success to form freshwater marsh.[40]

## III. PLANTING

## A. Site Preparation

Grading may be required to eliminate pockets of poor drainage in some situations. Sloping or filling may be necessary on eroding shorelines to provide a plantable slope. Even when plants can be established throughout much of the intertidal zone, as is the case on many of the Atlantic coast marshes, the potential width (landward to seaward) of a particular planting is dependent upon tidal amplitude and shore slope. Broader marshes can be established coincidentally with greater tidal ranges and more gradual sloping shorelines. The width of the beach at an elevation suitable for plant establishment will determine the relative stability of plantings made in areas subject to erosion. Waves are dampened as they pass through stands of marsh vegetation. The amount of dampening that occurs is directly related to the width of the marsh. From a survey of marsh plantings, Knutson[19] concluded that a practical minimum planting width is about 6.0 m in areas subject to erosion. If the potential planting area is not sufficiently wide, the shore must be graded to extend the planting area. Cultivation beyond that required to obtain a suitable planting width should be avoided. Substrates are typically more erodable once disturbed. Grading must be done long enough in advance to planting to allow for consolidation of the disturbed soil to take place. Otherwise, transplants may be dislodged by the first minor storm before sufficient anchoring roots develop.

Severe storms during the establishment period may dislodge young plants and seriously interfere with the establishment of the planting. This is particularly critical on direct seedings but is also often important on vegetative plantings. There are many exposed sites where marsh vegetation might do well once established. However, without some kind of temporary protection, new plantings may have little chance of establishing.

The solution of this type of problem will vary widely. On large deposits of sand lying in and above the intertidal zone, wind transport can be substantial. In such cases, properly placed sand fences may be used to protect plantings and prevent failure.[12] Movement by waves or currents is a frequent problem that is usually much more difficult to combat. Remedies that have been tried with varying success are breakwaters of scrap tires, baled hay, and scrap tires,[41] fiberglass,[14] and cloth or net mulch.[42] Unfortunately, temporary protection for a planting may cost more than the planting itself and may not be effective. There is a great need for imaginative development of temporary, inexpensive protective devices for this purpose.

Typically, the grasses, sedges, and rushes which form the intertidal marsh community are exposed to direct sunlight, because this zone lacks an overstory of either shrubs or trees. In general, woody vegetation is not found within the intertidal zone of the coastal U.S. The major exception to this is the occurrence of red mangroves in Southern Florida. On the Atlantic coast, marsh-elder (*Iva frutescens*) grows on the outer margin of salt marshes. It grows closer to the marsh than any other woody species and is appropriately referred to as the "High Tide Bush". However, erosion patterns may alter the normal zonation of plant communities on the shore. Progressive erosion may have obliterated the marsh and shrub transition zones leaving a mature forest overhanging an exposed bank. This condition will preclude the establishment of marsh vegetation because of its impact on availability of light. The area can be planted if all the overstory is cleared above the planting area and landward a distance of at least 3 to 5 m. Continued control of overstory species will be an essential part of maintenance of such sites.

## B. Fertilization

Nutrient supplies in regularly flooded salt marshes are usually adequate, particularly in established marshes in sediment-rich estuaries. Large quantities of nutrients are stored in fine-grained sediments, and this supply is regularly augmented by fresh deposits. However, fertilizers can be a useful tool in establishing new stands of marsh under certain circumstances. Freshly deposited or exposed sandy substrates are usually deficient in nutrients, particularly nitrogen and phosphorus. The addition of these nutrients will accelerate growth thereby shortening the time that establishing plants are most vulnerable to waves and currents.[11-14]

Fertilizer response is usually, but not always, confined to sandy substrates. Deficiencies acute enough to severely hamper plant establishment have been found on heavy-textured soils.[43] So far these appear to be confined to eroding shorelines, but this problem could be more widespread.

Demonstrated fertilizer response of salt marsh species has been to nitrogen and phosphorus. Benefits from additional potassium or micronutrients are unlikely under salt or brackish water conditions.

The form of nitrogen is important. Nitrate is subject to rapid denitrification under anaerobic conditions.[44]

Ammonia is utilized more efficiently by smooth cordgrass than the nitrate form, just the reverse from most upland plants.[13,45,46] This is probably true for other marsh species because ammonia is the normal form of nitrogen existing under anaerobic conditions. Results with urea have been similar to those with nitrate.[43]

Conventional materials such as ammonium sulfate and triple superphosphate are usually the most economical fertilizers for marsh plantings. Slow-release fertilizer such as Osmocote® and magnesium-ammonium-phosphate may be very effective on marsh plantings,[35] but there is no indication that these materials are better than conventional forms, properly used.[43] Slow-release materials may permit the use of lower rates, particularly of nitrogen, to obtain the same result and are more convenient in that they require fewer applications. However, they are far more expensive and where cost is a factor, as on large plantings, only the conventional forms appear to be practical.

Ammonium sulfate and triple superphosphate applied on the bare soil surface at low tide can be very effective and remain in place surprisingly well with little evidence of lateral movement.[12] However, surface application may be very ineffective on compact soils. Soluble as well as slow-release fertilizer materials should be applied in the planting holes or furrows and covered prior to reflooding on heavy-textured soils.

Split applications of nitrogen (ammonia forms) are likely to result in more efficient utilization than large single applications on sandy soils. Three applications of 30 to 50 kg of nitrogen (N) per hectare and one application of 30 to 50 kg of phosphate ($P_2O_5$) per hectare may be warranted during the first growing season on sands. The phosphate and the first nitrogen application should be made 2 to 4 weeks after planting or as soon as new growth appears. The other nitrogen application should follow at 6-week intervals.

A possible acceleration of eutrophication by the addition of fertilizers to estuaries should be considered. Although there are no data bearing directly on this problem, the judicial use of fertilizers in marsh establishment is unlikely to contribute significantly to the pollution load of most estuaries for the following reasons:

1.  Applied nitrogen utilization by marsh plantings can be quite efficient. Apparent recovery in aboveground growth in the year of application has been as high as 50%, comparable to that of upland crops.[13]
2.  The amount of nitrogen applied in a planting, encompassing only a small part of an estuary, is usually insignificant in comparison with the nitrogen regularly entering estuaries from other sources (agricultural, municipal, and industrial).
3.  Little fertilizer phosphorus is likely to leave a planted area because of the affinity of marsh sediments for this nutrient.
4.  Fertilization will normally be a one-season event, applied only in the year of establishment. The resulting marsh will be capable of immobilizing much larger quantities of pollutants in succeeding years.
5.  Slow-release material will probably contribute even less to the waters of the estuary.

## C. Planting Techniques

Seeds should be broadcast at low tide and covered 1 to 3 cm deep by tillage. It is usually advisable to till both before and after broadcasting to ensure more uniform coverage. Wet seeds will separate satisfactorily for broadcasting if mixed with dry sand. Seeding is usually feasible only in the upper 20 to 30% of the tidal range. Stands established by seeding will spread downslope by rhizome extension to the lower limit for the site. In general, seeding will be effective only on sites that are sheltered from wave activity.

The essentials in successful transplanting vegetative materials are to open a hole or

FIGURE 7.   Hand planting. (Courtesy of U.S. Army, Corps of Engineers, Coastal Eng. Res. Cent., Fort Belvoir, Va.)

furrow deep enough to accommodate the plant to the required depth, keep the hole open until the plant can be properly inserted to the full depth, close the opening, and firm the soil around the plant. This operation should be done during low water. It is virtually impossible to do a satisfactory job of transplanting while the surface is flooded. Openings close too rapidly and plants tend to float out. There are a number of tools and procedures that work well provided the substrate is exposed. Hand planting can be very satisfactory if adequate attention is given to details, particularly planting depth and soil firming after planting (Figure 7). It is the usual and most practical method for small scale plantings. Hand opening of planting holes is readily done with dibbles, spades, and shovels in soils that are not too compact. Portable power-driven augers work well in the more difficult soils. Normally planting crews work in pairs, one worker opening holes and the other inserting the plant and closing the hole. If fertilizer is added in the planting hole, this is done by a third worker who drops in a measured amount of material just after the hole is opened and before the plant is inserted.

Machine planting, where feasible, can do a much more uniform job and is far more economical than hand planting in large-scale plantings (Figure 8). Tractor-drawn planters designed to transplant crop plants such as cabbage, tomatoes, tobacco, etc. are available in most regions. Some may require alteration of the row opener for certain soils, but they can often be used without alteration. The principal barriers to machine planting will usually be (1) inadequate traction on compact substrates, (2) insufficient bearing capacity on soft sites, or (3) the presence of tree roots or stones that interfere with the functioning of the row opener.

Planting depth is not critical on sheltered sites. Most species do best planted 1 or 2 in. deeper than when dug or removed from pots. However, in planting exposed shores, it is often highly desirable to anticipate erosion or accretion trends that are likely to prevail during the first month or two after planting. Most species will develop satisfactorily when planted several inches deeper than normal, and where erosion is expected, it will be well to set them deeper. Where deposition is likely, they should be set very close to the depth they were growing when dug or when removed from pots.

## D. Plant Spacing

The spacing between planting hills is often an important consideration. The more closely plants are spaced, the more rapidly the marsh will achieve densities comparable

FIGURE 8.    Mechanical planting. (Courtesy of S.W. Broome.)

to natural marshes. However, small changes in plant spacing will greatly change the number of planting units required:

| | |
|---|---|
| 1 m spacing — | 10,000 planting units per hectare |
| ½ m spacing — | 40,000 planting units per hectare |
| ¼ m spacing — | 160,000 planting units per hectare |

Vegetative transplants (field-grown, plug, or peatpot) set on 1-m centers will, under average conditions, provide complete cover by early spring of the second growing season. Denser spacing (0.5 m) may be warranted on exposed sites or where early stabilization is required. Planting costs are in almost direct proportion to the number of plants planted. A 0.5 m planting will require four times as much. Differences between them will often not be distinguishable after the first growing season.

The optimum density for seeding appears to be around 100 viable seeds per square meter, but adequate stands have been obtained under favorable conditions with less than half this rate.

### E. Planting Season

Little information is available concerning optimal planting periods for species other than smooth cordgrass and saltmeadow cordgrass. Smooth cordgrass seeds germinate in nature rather early (late February or March, along the coast of the Carolinas; December and January in Florida). However, direct seeding has generally been more successful where seeds were held in cold storage and seeded in April or early May after storm hazards have diminished. Seedings made as late as June have established successfully. Smooth cordgrass vegetative material can be planted year round but not with equal success. Early spring planting avoids the winter storms and provides a longer growing season for establishment, particularly in the more northern latitudes. March, April, and early May probably represent the optimum planting season along the mid-Atlantic coast with the season starting somewhat later and becoming shorter northward. The practical planting season starts as early as February and extends much longer in the more southern extremes.

Saltmeadow cordgrass has a rather wide tolerance to time of planting, from late winter to early summer; Gallagher, Plumley, and Wolf[47] suggest autumn planting, but this has not been tested. Soil moisture content during and following planting is probably more important for this species than planting date.

## F. Planting Maintenance

Debris such as wood, styrofoam, algae, and dislodged submerged plants may accumulate in the planting areas and form a strand line. This material will smother and damage plantings particularly during the first two growing seasons. This litter should be removed in both the autumn and spring.

Canada and Snow geese are fond of the tender roots and rhizomes of marsh plants and may destroy a planted area before the plants are well established in areas of high winter wildfowl concentrations. Rope fences erected on the seaward edge of planted areas have been used successfully to exclude waterfowl during the first few growing seasons.

The fences consist of wood, metal, or plastic pickets strung with 1/8-in. nylon rope. The ropes are spaced at 15 cm (6-in.) intervals from the sediment surface to the water surface at high tide.[48]

Severe storms may cause damage to plantings particularly during the establishment period. Damaged areas should be replanted.

## IV. PLANTING COSTS

### A. Site Preparation

Sites which have little likelihood of being successfully stabilized can be modified to improve their suitability. The two primary methods of improving a site are (1) grading of the shoreline or (2) constructing a wave stilling device.

### 1. Grading the Shoreline

Grading the shoreline can increase the width of the area available for planting and the distance over which wave energy will be dissipated. This will usually improve the chances for successful plant establishment. The following is an estimate of the costs required to create a 1 vertical to 15 horizontal slope in front of a 1 m high bank.[49]

| Method | Cost/Cubic Meter | Cost/Linear Meter |
|---|---|---|
| Dredge (hydraulic placement) | $4.50 | $33.75 |
| Dump truck placement | 9.00 | 67.50 |

These values were calculated assuming: (1) there is easy access to the work site, (2) the fill material is within a reasonable distance to the work site, (3) all construction is in an area of low to moderate wave climate, (4) grading is included in the estimate, and (5) the hydraulic dredge is used for only large projects in excess of 1000 linear meters.

### 2. Wave Stilling Devices

Wave stilling devices are used to protect the planting from severe wave impact. These can be very helpful in protecting plants through the critical establishment period. The following is an estimate of constructing two, low-cost wave stilling devices.

| Method | Cost/Linear Meter |
|---|---|
| Rubber Tire breakwater (2 tires high, labor included*) | $25 (adapted from Webb and Dodd[16]) |
| Sandbag dike breakwater (1 m high with filter cloth included) | $100 (adapted from Eckert, Giles, and Smith[49]) |

* Labor is estimated at $30 per man-hour ($15 per man-hour direct costs plus 100% overhead).

## B. Harvesting, Processing, and Planting

The labor required to acquire or produce propagules and to plant is the principal cost of a project unless site preparation and/or temporary protection is required. Labor demands vary widely with species, availability of plants and seeds, type of propagule, accessibility of the site, soil type, size of operation, and degree of mechanization used.

Working hours in the intertidal zone are controlled by tidal regimes. Both harvesting and planting are usually confined to about a 5-hr period per tide. This restriction requires careful coordination for efficient operation and often adds substantially to the cost.

The most extensive experience is with sprigs (intact, single-stem plants) of smooth cordgrass. Estimates for digging, processing, and planting range from about 75 hills per man-hour for a manual operation[15] to about 200 hills per man-hour where digging and planting are mechanized.[12] Dodd and Webb[15] worked with a variety of species and found that the more difficult plants, giant reed and American bulrush, required about 1 man-hour per 35 hills.

Knutson and Woodhouse[21] have summarized labor requirements for planting vegetative materials based on existing literatures:

Sprigs: 10 man-hours per 1000 hills
Nursery Seedlings: 23 man-hours per 1000 hills
Plugs: 30 man-hours per 1000 hills

To estimate labor requirements for a particular project, first determine the number of planting units required as follows:

$$\text{Number of Planting Units} = \text{Area of Planting} \times \frac{1}{(\text{Plant Spacing})^2}$$

(Plant spacing is 0.5 m for sites subject to erosion and 1.0 m for sheltered sites.)

Second, determine the labor required to prepare and plant these units as follows:

$$\text{Labor Requires} = \text{Number of Planting Units} \times \frac{\text{Man-hours}}{1000 \text{ Planting Units}}$$

(As noted above, sprigs require about 10 man-hours per 1000 planting units, plugs about 30 man-hours per 1000 planting units, and nursery seedlings require about 23 man-hours per 1000 planting units.)

A typical project using sprigs spaced 1.0 m apart will require 10,000 planting units and 100 man-hours per hectare.

Seeding is much less labor intensive than planting vegetative materials. Most seeding projects will require only about 25 man-hours per hectare.

## C. Fertilization

The cost of fertilizer is variable but will probably cost no more than $150 to $250/ha (1980) including labor. Slow release fertilizer is more expensive, about $1500 to $2500/ha. However, the use of slow-release materials will eliminate the need for post-planting fertilizer applications.

## V. FACTORS CONTROLLING SUCCESS OR FAILURE

A number of factors are known to influence planting success. An early consideration in all potential plantings is the selection of species suitably adapted for the environment. Primary factors to consider are the plants natural elevational range with respect to tides and its natural salinity regime. These factors have been previously summarized for major species in Section II. Additional information on the environmental requirements for species not discussed in this chapter can be found in Kadlec and Wentz.[22]

Other factors influencing success or failure in marsh plantings are the soils and wave climate of the planting area. Following is a state-of-the-art discussion of these factors.

## A. Soils
### 1. Types

A few saltmarsh species are confined to certain soil types or conditions. However, most saltmarsh plants exhibit a wide tolerance of substrates. They may be found growing on mineral soils ranging from coarse sands to heavy clays and on peats and mucks of widely varying nutrient content and degree of decomposition. This does not mean that soils are unimportant to marsh establishment and growth.

Even under the most favorable conditions, transplants require several weeks to anchor themselves and still more time to develop appreciable protective effect. Stability of the substrate is essential to this process. Consequently, planting in loose sands is a poor risk if the site is likely to be subjected to substantial wave activity during this period. Even though net erosion may be minimal, substrate movement will usually dislodge the transplants before they can become fully anchored.

### 2. Salinity

Salinity is the one common factor that affects all salt marsh plants. These plants must have some salt tolerance, a prime requirement in this habitat. Some of the more tolerant species have the capacity to excrete salt through special structures (salt glands) in their leaves. A number of them possess another mechanism in their roots for screening toxic ions and slowing their inward penetration.[50]

Plants of the regularly flooded, low marshes, such as smooth cordgrass, are well equipped to live and grow in salinities up to $35^{\circ}/_{\circ\circ}$ (sea strength). However, these plants are usually quicker to establish and more productive in salinities below sea strength. Seeds and young seedlings are usually more sensitive to salt concentration than are established plants.

Toxic concentrations usually do not develop in sandy marsh soils within the regularly flooded zone. The salinity in such soils tends to remain close to that of the surrounding water. This may not always be true of fine-textured soils in which salt may accumulate through ion exclusion by roots,[51] although it does not appear to be a common occurrence in natural marshes. Salt accumulation in the fine-textured marsh soils is probably held to a minimum by the drainage normally provided by root channels and animal burrows.

Salt damage may occur on newly planted areas due to concentrations through evaporation in the zone between neap tide high water and spring tide high water during periods of low rainfall and warm temperatures following spring tides. This also occurs in sounds and bays subject to wind setup in which the wind pattern results in extended periods of low water during hot weather, as in Core Sound, N.C. Under these conditions, soil-water salinities of 50 to 75 $^{\circ}/_{\circ\circ}$ may develop and persist until diluted by rainfall or tidal inundation.[12]

Irregularly flooded, high marshes are subject to occasional salt buildup through evaporation and ion exclusion regardless of soil texture. However, this is usually limited to poorly drained areas that are flooded by storm tides. In humid climates precipitation plus fresh water seepage from higher ground tends to keep salinities in most high marshes well below sea strength. Under more arid conditions, salt concentrations often exclude marsh species altogether.

In general, suitable plants may be found which can be established in salinities up to about sea strength (35 $^o/_{oo}$). In bays and estuaries where salinities seasonally exceed sea strength, marsh planting is not likely to succeed. If salinity is a suspected problem, the presence, abundance, and vitality of native intertidal plants in sheltered areas in the vicinity of the proposed project will be an indicator of probable success.

### B. Wave Climate
#### 1. Indicators of Wave Severity
In brackish and salt water areas, wave climate severity has a major influence on marsh establishment. Three shoreline characteristics — fetch, shore configuration, and sediment grain size — are useful indicators of wave climate severity and planting success.[19] *Fetch*, the distance the wind blows over water to generate waves, is inversely related to successful erosion control. *Shore configuration*, the shape of the shoreline, is a subjective measure of the shoreline's vulnerability to wave attack. For example a cove is relatively sheltered while a headland is vulnerable to wave attack from many directions. The *grain size* of beach sands are also related to wave energy. Fine-grained sands frequently indicate higher energy beaches. Two additional factors should be considered when evaluating wave climate: boat traffic and offshore depth. Shore areas in close proximity to boat traffic will be subject to ship-generated waves. Shallow offshore depths impede the growth and development of larger waves. However, no method is available for numerically evaluating boat traffic and offshore depth.

#### 2. Method for Evaluating Wave Climate
Knutson et al.[19,21] developed a method for evaluating wave climate based upon observed relationships between fetch, shore configuration, and grain size and success in controlling erosion in 86 salt marsh plantings in 12 coastal states. The method uses a Vegetative Stabilization Site Evaluation Form (Figure 9) to evaluate planting potential on a case-by-case basis. The user (1) measures each of four *Shore Characteristics* for the area in question, (2) identifies the Descriptive Categories which best describe the area, (3) notes the weighted score associated with each Descriptive Category in Column 3, (4) calculates a *Cumulative Score,* and (5) notes the success rate associated with the appropriate range of Cumulative Scores under *Score Interpretation.* Sites with a Cumulative Score of 300 or greater (observed success rate of 100%) are very promising planting environments. However, even sites with a Cumulative Score of 201 to 300 (observed success rate of 50%) will often constitute an acceptable risk.

## VI. HABITAT VALUE OF PLANTED MARSH

Salt marshes are valued as sources of primary production (energy) and as nursery grounds for sport and commercial fishery species; as a system for storing and recycling nutrients and pollutants such as nitrogen, phosphorus, and heavy metals and as a natural buffer to shore erosion. Once established, marsh plantings function as natural salt marshes and gradually develop comparable animal populations.[52,53]

### A. Marsh Ecology
Little of the biomass of salt marsh, about 5%, is consumed while the plant material is still living. Grasshoppers and plant hoppers graze on the grass and are in turn eaten by spiders and birds. Direct consumption of rhizomes and culms of marsh grasses by water fowl may be significant locally near wintering grounds. Periwinkles graze on algae growing on the grass. The majority of the energy is believed to move through the detrital food chain. Dead grass is broken down by bacteria in the surrounding waters and on the surface of the marsh. This process greatly decreases the total energy

| 1. SHORE CHARACTERISTICS | 2. DESCRIPTIVE CATEGORIES (SCORE WEIGHTED BY PERCENT SUCCESSFUL) | | | | 3. WEIGHTED SCORE |
|---|---|---|---|---|---|
| **a. FETCH-AVERAGE** AVERAGE DISTANCE IN KILOMETERS (MILES) OF OPEN WATER MEASURED PERPENDICULAR TO THE SHORE AND 45° EITHER SIDE OF PERPENDICULAR | LESS THAN 1.0 (0.6) | 1.1 (0.7) to 3.0 (1.9) | 3.1 (1.9) to 9.0 (5.6) | GREATER THAN 9.0 (5.6) | |
| | (87) | (66) | (44) | (37) | |
| **b. FETCH-LONGEST** LONGEST DISTANCE IN KILOMETERS (MILES) OF OPEN WATER MEASURED PERPENDICULAR TO THE SHORE OR 45° EITHER SIDE OF PERPENDICULAR | LESS THAN 2.0 (1.2) | 2.1 (1.3) to 6.0 (3.7) | 6.1 (3.8) to 18.0 (11.2) | GREATER THAN 18.0 (11.2) | |
| | (89) | (67) | (41) | (17) | |
| **c. SHORELINE GEOMETRY** GENERAL SHAPE OF THE SHORELINE AT THE POINT OF INTEREST PLUS 200 METERS (662 FT) ON EITHER SIDE | COVE (85) | MEANDER OR STRAIGHT (62) | HEADLAND (50) | | |
| **d. SEDIMENT** GRAIN SIZE OF SEDIMENTS IN SWASH ZONE (mm) | 0.0 – 0.4 (84) | 0.4 – 0.8 (41) | 0.8 – or greater (18) | | |
| **4. CUMULATIVE SCORE** | | | | | |
| **5. SCORE INTERPRETATION** | | | | | |
| **a. CUMULATIVE SCORE** | 0 – 200 | 201 – 300 | 300 – or greater | | |
| **b. SUCCESS RATE** | 15 % | 50 % | 100 % | | |

FIGURE 9. Vegetative stabilization site evaluation form.[19,21]

but increases the concentration of protein, thereby increasing the food value. Some detrital particles and mud algae are eaten by a variety of detritus feeders such as fiddler crabs, snails, and mussels; these organisms are, in turn, eaten by mud crabs, rails, and racoons. The remaining detritus, augmented by the dead matter from the primary and secondary consumers, is washed from the marsh by tidal action as new export. This exported detritus with material from submergent aquatic plants and the plankton, feeds the myriad of larvae and mature fish and shellfish which use estuaries, bays, and adjoining shallow waters. Marsh grasses may account for most of the primary production of the system in waters where high turbidity reduces light penetration, thereby reducing phytoplankton and submergent aquatic production.

The productivity and utilization of high marsh has received less attention that that of low marshes. Indications are that net production of some high marsh may equal that of many low marshes. The important difference, however, is that the export mechanism of frequent tidal flushing is absent in high marsh. Consequently, much of the high marsh biomass goes into peat formation rather than into the estuarine food chain. For this reason, high marsh appears to be of much less direct value to the estuary. However, the high marsh is used much more extensively for resting and nesting by shorebirds and waterfowl.

The rigorous environment of the salt marsh sharply limits the number of animals that live there. These areas are used by birds such as herons, rails, sandpipers, geese, ducks, and songbirds and by raccoons. A large population of animals live in or on the mud surface. The more conspicuous are fiddler crabs, mussels, clams, and periwinkles. Less obvious, but more numerous are annelid and obligochaete worms and insect larvae. In addition, larvae, juveniles, and adults of many shellfish and fish are commonly found in the marsh creeks.

FIGURE 10.    Oldest reported saltmarsh planting in U.S.[19]

## B. Nutrient Cycling

Salt marshes have substantial absorptive capacities for potential pollutants such as nitrogen, phosphorus, and heavy metals,[8,12] Increased growth of salt marsh species, particularly smooth cordgrass in response to nutrients has been noted at several locations.[12,14,46,54,55] Under some circumstances, smooth cordgrass will increase growth in response to applications of as much as 672 kg of nitrogen and 74 kg of phosphorus per hectare per year as fertilizer.[12,14] Apparent utilization of applied nitrogen may be as high as 40 to 60% in shoot growth alone, a value that compares favorably with upland field crops. The potential for substantial recycling and exporting of nutrients to the estuary exists.

The absorption, conversion, and recycling capabilities of marsh plants offer real opportunities for water purification.[56]

## C. Erosion Control

Erosion control is probably the most common objective of marsh planting projects. Establishing a fringe marsh on an eroding shoreline is a very low cost, environmentally beneficial alternative to structural shore protection.

Marsh plants perform two functions in abating erosion. First, their aerial parts form a flexible mass which dissipates wave energy. As wave energy is diminished, the offshore and longshore transport of sediment is reduced. Optimally, dense stands of marsh vegetation can create a depositional environment, causing accretion rather than erosion of the shore. Second, many marsh plants form dense root-rhizome mats which add stability to the shore sediment. This protective mat is of particular importance during severe winter storms when the aerial stems provide only limited resistance to the impact of waves.

This erosion control alternative has been used successfully for many years in the U.S. In the winter of 1928, a property owner on the eastern shore of the Chesapeake Bay planted smooth cordgrass along more than 1 km of shoreline in an attempt to reduce erosion. This shoreline has remained stable for more than 50 years and is the oldest reported example of shore stabilization with salt marsh in the U.S.[19] (Figure 10). Similarly, in 1946, a landowner on the Rappahanock River in Virginia graded an eroding shoreline and planted several varieties of salt tolerant plants. This planting has prevented erosion for more than 30 years.[18,57]

## VII. SUMMARY

A.   Principle Species
  1.   Smooth Cordgrass (*Spartina alterniflora*)
      a.   Elevational range — mean low water to MHW where tidal range less than 2.0 m; mean tide to MHW where tidal range greater than 3.0 m.
      b.   Salinity range — 5 to 35 $^o/_{oo}$.
  2.   Saltmeadow Cordgrass (*Spartina patens*)
      a.   Elevational range — MHW to estimated highest tide (occasionally flooded).
      b.   Salinity range — optimal range 0 to 15 $^o/_{oo}$ but will grow next to smooth cordgrass marshes of 35 $^o/_{oo}$.
  3.   Black Needle Rush (*Juncus roemerianus*)
      a.   Elevational range — MHW to estimated highest tide (occasionally flooded).
      b.   Salinity range — less than 20 $^o/_{oo}$.
  4.   Saltgrass (*Distichlis spicata*)
      a.   Elevational range — poorly drained saline patches above MHW.
      b.   Salinity range — up to about 40 $^o/_{oo}$.
  5.   Big Cordgrass (*Spartina cynosuroides*)
      a.   Elevational range — above MHW.
      b.   Salinity range — Up to about 20 $^o/_{oo}$.
  6.   Common Reed
      a.   Elevational range — above mean high tide.
      b.   Salinity range — 0 to 20 $^o/_{oo}$.
B.   Plant Materials
  1.   Seeds — least expensive planting method but limited to very sheltered sites.
  2.   Sprigs — least expensive vegetative material to obtain and easy to handle, transport, and plant.
  3.   Pot-grown Nursery Seedlings — more expensive to grow and plant, more awkward to handle and transport, but relatively easy to produce.
  4.   Plugs — most expensive to obtain, difficult to transport and plant, only used when other sources not available.
C.   Planting Methods
  1.   Hand Planting (Dibbles, Spades, and Shovels) — suitable for all plant materials.
  2.   Power-driven Auger — useful for difficult soils and for pot-grown seedlings and plugs.
  3.   Machine Planting (Cabbage, Tomato, and Tobacco Planters) — very efficient for large-scale plantings of sprigs.

## VIII. RESEARCH NEEDS

Most coastal marsh species are hardy and aggressive colonizers. Though research is lacking on many species, attempts to establish new marshes typically meet with success. The question for researchers is no longer "if marshes can be restored". Attention now must be focused upon developing improved, cost-effective planting techniques. Today, marshes planted under contract cost from $15,000 to $30,000/ha. This is an order of magnitude more costly than other agricultural activities.

Marsh planting represents an important alternative to structural methods of erosion control. However, considerable research is needed before vegetative stabilization will be directly comparable to engineering solutions. Research on shore protection must address what is now the major obstacle to plant establishment — wave action. To date, there is little evidence concerning the likelihood of planting success under different wave conditions. It is not even clear which planting technique is most effective in the face of severe wave conditions.

# REFERENCES

1. Gosselink, J. G. and Baumann, R. H., Wetland inventories: wetland loss along the United States coast, *Z. Geomorphol.,* 34, 173, 1980.
2. Macomber, R. H., Salt marshes, *Bull. Am. Littoral Soc.,* Special Issue: The Coast, 12(1), 1, 1979.
3. Shaw, S. P. and Fredine, C. G., Wetlands of the United States, their extent and their value to water-fowl and other wildlife, Circ. 39, Fish and Wildlife Service, U.S. Department of the Interior, 1956.
4. Odum, E. P., The role of tidal marshes in estuarine production, *N.Y. State Conservat.,* 15(6), 12, 1961.
5. Teal, J. M., Energy flow in the salt marsh ecosystem of Georgia, *Ecology,* 43(4), 614, 1962.
6. Odum, E. P. and de la Cruz, A. A., Particulate organic detritus in a Georgia salt marsh-estuarine ecosystem, *Estuaries,* Lauff, G. H., Ed., American Association for the Advancement of Science, Washington, D.C., 1967, 383.
7. Cooper, A. W., Salt marshes, *Coastal Ecological Systems of the United States,* Odum, M. T., Copeland, B. J., and McNamon, E. F., Eds., Institute of Marine Science, University of North Carolina, Chapel Hill, 1969, 567.
8. Williams, R. B. and Murdock, M. B., The potential importance of *Spartina alterniflora* in conveying zinc, manganese, and iron into estuarine food chains, Proc. 2nd Natl. Symp. Radioecol., U.S. Atomic Energy Commission Conf., 670503, Ann Arbor, Mich., 1969.
9. Coates, D. R., *Environmental Geomorphology and Landscape Conservation,* Vol. 1, Benchmark Papers in Geology, Dowden, Hutchinson and Rose, Stroudsburg, Pa., 1972.
10. American Littoral Society, Protecting wetlands — what you should know, Pamphlet, American Littoral Society, Sandy Hook, Highlands, N.J., 1981.
11. Woodhouse, W. W., Jr., Seneca, E. D., and Broome, S. W., Marsh building with dredge spoil in North Carolina, Bull. 445, Agric. Exp. Stn., North Carolina State University, Raleigh, July 1972.
12. Woodhouse, W. W., Jr., Seneca, E. D., and Broome, S. W., Propagation of *Spartina alterniflora* for substrate stabilization and salt marsh development, TM 46, U.S. Army, Corps of Engineers, Coastal Eng. Res. Cent., Fort Belvoir, Va., August 1974.
13. Woodhouse, W. W., Jr., Seneca, E. D., and Broome, S. W., Propagation and use of *Spartina alterniflora* for shoreline erosion abatement, TR 76-2, U.S. Army, Corps of Engineers, Coastal Eng. Res. Cent., Fort Belvoir, Va., August 1976.
14. Garbisch, E. W., Jr., Woller, P. B., and McCallum, R. J., Salt marsh establishment and development, TM 52, U.S. Army, Corps of Engineers, Coastal Eng. Res. Cent., Fort Belvoir, Va., June 1975.
15. Dodd, J. D. and Webb, J. W., Establishment of vegetation for shoreline stabilization in Galveston Bay, MP 6-75, U.S. Army, Corps of Engineers, Coastal Eng. Res. Cent. Fort Belvoir, Va., 1975.
16. Webb, J. W. and Dodd, J. D., Vegetation establishment and shoreline stabilization, Galveston Bay, Texas, TP 76-13, U.S. Army, Corps of Engineers, Coastal Eng. Res. Cent., Fort Belvoir, Va., August 1976.
17. U.S. Army Waterways Experiment Station, Wetland habitat development with dredged material: engineering and plant propagation, TR DS-78-16, U.S. Army, Corps of Engineers, Waterways Exp. Stn., Vicksburg, Miss., December 1978.
18. Sharp, W. C. and Vaden, J., Ten-year report on sloping techniques used to stabilize eroding tidal river banks, *Shore Beach,* 38, 31, 1970.
19. Knutson, P. L., Ford, J. C., Inskeep, M. R., and Oyler, J., National survey of planted salt marshes (vegetative stabilization and wave stress), *Wetlands,* 1, 129, 1981.

20. **Woodhouse, W. W., Jr.,** Building salt marshes along the coasts of the continental United States, SR 4, U.S. Army, Corps of Engineers, Coastal Eng. Res. Cent., Fort Belvoir, Va., May 1979.

21. **Knutson, P. L. and Woodhouse, W. W.,** Shore stabilization with salt marsh vegetation, SR, U.S. Army, Corps of Engineers, Coastal Eng. Res. Cent., Fort Belvoir, Va., in press.

22. **Kadlec, J. H. and Wentz, W. A.,** State of the art survey and evaluation of marsh plant establishment techniques: induced and natural, Vol. 1, TR D-74-9, U.S. Army, Corps of Engineers, Waterways Exp. Stn., Vicksburg, Miss., 1974.

23. **Teal, J. M. and Kanwisher, J. W.,** Gas transport in the marsh grass *Spartina alterniflora, J. Exp. Bot.,* 17(51), 355, 1966.

24. **Anderson, C. E.,** A review of structure in several North Carolina salt marsh plants, *Ecology of Halophytes,* Academic Press, New York, 1974, 303.

25. **Chapman, V. J.,** *Salt Marshes and Salt Deserts of the World,* Leonard Hill Ltd., London, 1960.

26. **Stalter, R. and Batson, J.,** Transplantation of salt marsh vegetation, Georgetown, South Carolina, *Ecology,* 50, 1087, 1969.

27. **Mooring, M. T., Cooper, A. W., and Seneca, E. D.,** Seed germination response and evidence for height ecophenes in *Spartina alterniflora* from North Carolina, *Am. J. Bot.,* 58(1), 48, 1971.

28. **Shea, M. L., Warren, R. S., and Niering, W. A.,** The ecotype-ecophene status of varying height forms of *Spartina alterniflora, Bull. Ecol. Soc. Am.,* 53(2), 15, 1972.

29. **Seneca, E. D.,** Germination and seedling response of Atlantic and Gulf Coasts populations of *Spartina alterniflora, Am. J. Bot.,* 61(9), 947, 1974.

30. **Newton, N. H.,** personal communication, 1976.

31. **Terry, W. O., Udess, H. F., and Zarudsky, J. D.,** Tidal marsh restoration at Hempstead, Long Island, *Shore Beach,* 42(2), 36, 1974.

32. **Knutson, P. L.,** Designing for bank erosion control with vegetation, *Coastal Sediments 77,* Proc. Am. Soc. Civil Eng., November 1977, 716.

33. **Banner, A.,** Revegetation and maturation of a restored shoreline in the Indian River, Florida, *Proc. 4th Annu. Conf. Restorat. Coastal Vegetation Florida,* Lewis, R. R., III and Cole, G. P., Eds., Hillsborough Community College, Tampa, Fla., May 1977, 13.

34. **Hoagland, D. R. and Arnon, D. I.,** The water-culture method for growing plants without soil, Circ. 347, Calif. Agric. Exp. Stn., Berkeley, December 1938.

35. **Garbisch, E. W., Jr.,** Recent and planned marsh establishment work throughout the contiguous United States, a survey and basic guidelines, CR D-77-3, U.S. Army, Corps of Engineers, Waterways Exp. Stn., Vicksburg, Miss., April 1977.

36. **Hurme, A. K.,** unpublished data, 1980.

37. **Chabreck, R. H.,** Vegetation, water and soil characteristics of the Louisiana coastal region, Bull. No. 664, Louisiana Agric. Exp. Stn., Baton Rouge, La., September 1972.

38. **Hardisky, M. H. and Riemold, R. J.,** Buttermilk Sound habitat development site, Glynn County, Georgia, Georgia Department of Natural Resources, 1979.

39. **Restick, S. S., Frederick, S. W., and Buckley, E. H.,** Transplants of *Typha* and the distribution of vegetation and algae in a reclaimed estuarine marsh, *Bull. Torrey Bot. Club.,* 103, 157, 1976.

40. **Garbisch, E. W., Jr. and Coleman, L. B.,** Tidal freshwater marsh establishment in upper Chesapeake Bay: *Pontedera cordata* and *Peltandra virginica, Freshwater Wetlands Ecological Processes and Management Potential,* Good, R. E., Whigham, D. E., and Simpson, R. L., Eds., Academic Press, New York, 1978, 285.

41. **Broome, S. W.,** personal communication, 1978.

42. **Morris, J. H., Newcombe, C. L., Huffman, R. T., and Wilson, J. S.,** Habitat development field investigations, Salt Pond No. 3 marsh development site, South San Francisco Bay, Calif., summary report, TR D-78-57, U.S. Army, Corps of Engineers, Waterways Exp. Stn., Vicksburg, Miss., 1978.

43. **Broome, S. W., Seneca, E. D., and Woodhouse, W. W., Jr.,** The effects of source, rate, and placement of N and P fertilizers on growth of *Spartina alterniflora* transplants in North Carolina, *Estuaries,* in press.

44. **Patrick, W. H., Jr. and Mahapatra, I. C.,** Transformation and availability to rice of nitrogen and phosphorus in water-logged soils, *Adv. Agron.,* 20, 323, 1968.

45. **Gosselink, J. G.,** Growth of *Spartina patens* and *S. alterniflora* as influenced by salinity and source of nitrogen, *Coastal Studies Bulletin No. 5,* Louisiana State University Center for Wetland Resources, Baton Rouge, La., 1970, 97.

46. **Mendelsohn, J. A.,** The influence of N level, form, and application method on growth response of *Spartina alterniflora* in North Carolina, Botany Department, North Carolina State University, Raleigh, 1978.

47. **Gallagher, J. L., Plumley, F. G., and Wolf, P. L.,** Underground biomass dynamics and substrate selective properties of Atlantic coastal salt marsh plants, TR D-77-28, U.S. Army, Corps of Engineers, Waterways Exp. Stn., Vicksburg, Miss., December 1977.

48. Garbisch, E. W., Jr., personal communication, 1977.
49. Eckert, J. W., Giles, M. L., and Smith, G. M., Design concepts for in-water containment structures for marsh habitat development, TR D-78-31, U.S. Army, Corps of Engineers, Waterways Exp. Stn., Vicksburg, Miss., 1978.
50. Waisel, Y., *Biology of Halophytes,* Academic Press, New York, 1972.
51. Smart, R. M. and Barko, J. W., Influence of sediment salinity and nutrients on the physiological ecology of selected salt marsh plants, *Estuarine Coastal Mar. Sci.,* 7, 487, 1978.
52. Cammen, L. M., Microinvertebrate colonization of *Spartina* marsh artificially established on dredge spoil, *Estuarine Coastal Mar. Sci.,* 4(4), 357, 1976.
53. Cammen, L. M., Seneca, E. D., and Copeland, B. J., Animal colonization of salt marshes artificially established on dredge spoil, TP 76-7, U.S. Army, Corps of Engineers, Coastal Eng. Res. Cent., Fort Belvoir, Va., June 1976.
54. Valiela, I. and Teal, J. M., Nutrient limitation in salt marsh vegetation, *Ecology of Halophytes,* Reimold, R. J., and Queen, W. H., Eds., Academic Press, New York, 1974, 547.
55. Patrick, W. H., Jr. and Delaune, R. D., Nitrogen and phosphorus utilization by *Spartina alterniflora* in a salt marsh in Barataria Bay, Louisiana, *Estuarine Coastal Mar. Sci.,* 4, 59, 1976.
56. Woodhill, G. M., Recycling sewage through plant communities, *Am. Sci.,* 65, 556, 1977.
57. Phillips, W. A. and Eastman, F. D., Riverbank stabilization in Virginia, *J. Soil Water Conservat.,* 14, 257, 1959.

Chapter 3

# SALT MARSHES OF THE NORTHEASTERN GULF OF MEXICO

## William L. Kruczynski

## TABLE OF CONTENTS

# I. INTRODUCTION

Salt marshes are herbaceous plant communities which grow in the intertidal zone along margins of estuaries. They normally occur where the shoreline slope is gradual and the wave energy is low. Estuaries are located at the mouths of rivers; in bays, bayous, and sounds; and the estuarine area of the Gulf of Mexico is the largest in the U.S., excluding Alaska. There are 207 individual estuaries that exist along the Gulf Coast and most of the $24.9 \times 10^3$ km of estuarine shoreline is vegetated by salt marshes, mangrove forests, and submerged seagrass beds. This chapter will review the characteristics of those natural marshes which exist along the Gulf coastline from Tarpon Springs (Pinellas County), Fla. through Mississippi, and summarize studies on the establishment of marsh communities on dredged spoil.

## A. Existing Tidal Wetlands

Salt marsh vegetation of the eastern Gulf of Mexico consists predominantly of *Spartina alterniflora, S. patens, Distichlis spicata, Juncus roemerianus,* and a few species of succulent dicotyledons.[1] Mangroves, predominantly *Avicennia germinanis* and *Laguncularia racemosa,* range northward along the Gulf Coast of Florida to about Cedar Key (Levy County). Mangroves are not dominant in intertidal wetlands of the majority of the geographic area discussed in this chapter. Generally, tidal marshes are vegetated by few plant species. Plant species diversity increases from estuarine marshes toward freshwater wetlands. Similarly, plant diversity increases from the edge of the water toward the upland tree line.[2,3]

The total area of tidal marsh along the entire Gulf coast is about $2.4 \times 10^6$ ha, or approximately 65% of all the tidal marshes of the U.S.[4] Louisiana contains most of the Gulf's tidal marsh area (64%), followed by Texas (16%), Florida (9%), Mississippi (1%), and Alabama (0.6%).[5] Tidal marshes of the area covered by this chapter are predominantly *Juncus roemerianus* systems. That species is dominant in approximately 30% of the tidal marsh area in Florida (entire state), 51% of the tidal marsh area in Alabama, and 92% of tidal marshes in Mississippi (Table 1).[6] Approximately $1.0 \times 10^5$ Ha of tidal marsh exist along the western coast of Florida from Tarpon Springs northward through Escambia County. Of these approximately $0.6 \times 10^5$ (60%) are *Juncus* dominated systems.

## B. Plant Zonation

The most striking feature of salt marshes is the zonation of plant species. Zonation is largely controlled by salinity and range of tidal flooding, but also by physical and chemical parameters of the substrate.[1,2,7-9] A cross-section of a typical salt marsh of the northeastern Gulf is shown in Figure 1. *Spartina alterniflora* occurs as a pure stand along the outer edge of the marsh. This species exists from below sea level to the elevation of the highest predicted tide, which is approximately 0.8 m at Apalachee Bay, Fla.[10] Although it may occur throughout this tidal range, *S. alterniflora* is densest and most productive nearer the open water.[11] *Juncus roemerianus* grows in vast stands landward of the creekside *S. alterniflora* zone. Culms of *Juncus* in a typical marsh are so dense that horizontal movement of tidal water into the marsh is slowed. A thick, peat-like soil forms beneath a *Juncus* marsh, which is less permeable to water than that of any other salt marsh community.[1] The upper edge of the *Juncus* zone may be marked by the salt barren, or salt flat, an area where high soil salinity limits plant growth to a few well adapted species, such as *Salicornia bigelovii, S. virginica,* and *Batis maritima. Distichlis spicata* and/or *Spartina patens* may exist above the salt barrens or the upper limits of the *Juncus* zone. These species are normally inundated by storm tides. This zone, as well as the others, may vary in width from a few feet to

## Table 1
## AREA OF TIDAL MARSHES, *JUNCUS* MARSHES, AND TIDAL MARSH LOST 1947—1967

|  | Tidal marshes (ha × 10⁵) | *Juncus* marshes (ha × 10⁵) | % *Juncus* | Tidal marsh lost 1947—1976 ha |
|---|---|---|---|---|
| Florida (total) | 2.92 | 0.89 | 30 | 24,170 |
| Florida (Atlantic) | 0.89 | 0.28 | 31 | — |
| Florida (Gulf) | 2.02 | 0.61 | 28 | — |
| Alabama | 0.12 | 0.08 | 66 | 810 |
| Mississippi | 0.28 | 0.24 | 92 | 688 |

Data from References 6 and 28.

hundreds of feet depending upon the slope of the marsh. Normally, a shrub zone is found between the edge of the marsh and the area vegetated by upland trees. Common shrubs found in the Gulf coast are *Myrica cerifera, Baccharis halimifolia,* and *Iva frutescens.*

## C. Productivity

Tidal marshes are among the most productive natural ecosystems in the world. Most of the ecological research on salt marshes has concerned their role as producers of organic matter which decomposes into detritus. Detritus forms the basis of the food web in estuarine and marine environments. Macrophyte production represents approximately 75% of the total plant production in the estuarine-wetland complex of the Gulf coast.[5] Geographic variation in salt marsh macrophyte production was reviewed by Turner,[12] who found that there is considerable variation in estimates of net primary production within and among marshes. Estimates range from approximately 300 to 2000 g dry weight per square meter per year and increase with decrease in latitude. Turnover of dead material is dependent on physical forces, such as tidal amplitude and, in general, seems to increase with temperature.[12] Kruczynski et al.[13] found the net aerial production of *J. roemerianus* in a north Florida marsh decreased landward from 949 g/m³/year in low marsh to 595 g/m²/year in the upper marsh. Height and diameter of leaves and diameter of rhizomes also decreased landward. It has been recently shown by Mendelsson et al.[14] that the degree of soil reduction is responsible for the different growth forms known to exist in *Spartina alterniflora.* Metabolic adaptation to differing rates of flooding has resulted in clearly distinguishable tall and short *Spartina* stands. Perhaps, the different growth forms of *Juncus* observed by Kruczynski et al.[13] are due to a similar phenomenon. Gabriel and de la Cruz[15] found that a marsh community in Mississippi, dominated by *J. roemerianus,* produced approximately 1108 g/m²/year. Gosselink et al.[16] studies the productivity of seven marsh plant species on the Gulf coast over a 2-year period. They found, as Kruczynski et al.[13] also reported, that the harvest technique of Weigert and Evans[17] yields the most consistent, reliable estimates of net productivity. Gosselink et al.[16] reported net productivity (g/m²/year) to be *Spartina patens* — 4200; *Juncus roemerianus* — 3300; *Distichlis spicata* — 2900; *Phragmites communis* — 2400; *Sagittaria falcata* — 2300; *Spartina cynosuroides* — 1100. In this study, fresh and brackish marsh species supported high levels of productivity even though they did not receive the tidal subsidy which occurs in salt marshes.

## D. Animal Communities

The monotypic stands characterizing the lower marsh areas have been shown to be excellent habitat which supports diverse benthic and fish communities. Subrahmanyam

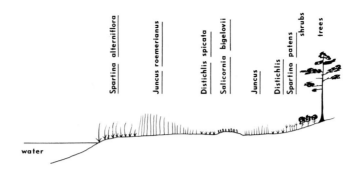

FIGURE 1.    Plant zonation of a typical North Florida tidal marsh.

et al.[18] reported 51 species of benthic invertebrates common in *Juncus* marshes of the Florida coast. Zonation of animals occurred, with lower marsh areas being populated by *Littorina, Cyathura,* tanaidaceans, and polychaetes, while high marsh areas were populated by *Uca, Melanpus,* and *Cerithidea.* Mean density of marsh invertebrates was 475/m² during a 12-month study. The relationship of marine macroinvertebrates to salt marsh plants was reviewed by Kraeuter and Wolf.[19]

There were 53 species of fish found to occur in a north Florida *Juncus* marsh,[20] with *Fundulis similis, F. grandis, Cyprinodon variagatus, Aedinia xenica, Poecilia latipina,* and *Menidia beryllina* being the most common marsh species. The role of tidal marshes as nursery habitat for many commercially important species is well-documented.[5,20-25] Approximately 48% of the 103 species of finfish and shellfish which make up the bulk of the commercial fishery organisms and 45% of the 60 recreational finfish species in the Gulf of Mexico are estuarine dependent.[5] These estuarine-dependent species comprise 90 to 97% of commercial biomass landed and 80% of the recreational biomass taken in the Gulf.[26]

### E. Wetland Destruction

When wetlands are filled, drained, paved, diked, or otherwise altered, a vital link in the marine food chain is damaged or destroyed. Precise data on the total acres of wetlands lost to dredging and filling are not available since wetland definitions are only now being standardized and our original wetland heritage was never fully mapped or inventoried. Precise data on existing wetlands will be available with the completion of the National Wetland Inventory which is currently being conducted by the U.S. Fish and Wildlife Service. The most commonly accepted estimate of total national wetlands which have been lost is between 30 and 40%. However, percentages are calculated from surveys designed for different purposes.[27] The U.S. Environmental Protection Agency (EPA)[28] summarized losses of coastal wetlands by dredging and filling from 1947 to 1967, based on data presented by Shaw and Fredine.[29] Between that period, Florida lost 24,200 ha, Georgia 324 ha, Alabama 810 ha, and Mississippi 688 ha. These data are probably low estimates, since a detailed study of Mississippi shows losses there to be much higher. Eleuterius[2] estimated that, in Mississippi, approximately 400 ha of marsh were filled prior to 1930, mostly for highway construction and residential and industrial development around cities such as Biloxi, Pascagoula, and Ocean Springs. Since 1930, approximately 3308 ha of salt marsh have been filled for industrial and housing developments. There have been 34 ha of marsh used as garbage dumps and subsequently filled. The result of these filling activities was that production of Mississippi marshes was reduced by 8.4 million t (dry weight) annually by 1930 and presently 72 million t of organic production are annually lost due to filling of productive coastal marshes.

## Table 2
### SITES OF CORPS OF ENGINEERS ORDERED FILL REMOVAL IN NORTHWESTERN FLORIDA

| Violation | County | Type wetland | Acreage (ha) | Date restored | Action |
|---|---|---|---|---|---|
| George Hunt | Pasco | *Juncus-Spartina* | 0.2 | 1977 | No planting |
| Pantori | Citrus | *Typha-Cladium* | 1.6 | 1976 | No planting |
| Butler | Citrus | Hardwood | 0.2 | 1980 | No planting |
| Weisman | Citrus | Hardwood | 0.6 | 1980 | Maple, cypress, gum |
| Suncoast City | Citrus | *Juncus-Spartina* | 2.0 | 1978 | No planting |
| Carlton McKinney | Levy | *Juncus* | 0.4 | 1979 | No planting |
| Levy County Commissioners | Levy | *Distichlis-Juncus* | 0.4 | 1975 | *Spartina* seeds |
| E. J. James | Taylor | *Juncus-Spartina* | 0.2 | 1975 | |
| Bingham and Bond | Bay | *Juncus* | 0.4 | 1979 | No planting |
| Eastern Marine | Bay | *Spartina* | 0.2 | 1981 | *Spartina* sprigs |
| Eastern Marine | Bay | *Spartina patens, Cludium, Juncus* | 0.2 | 1981 | Sprigs |
| Naval Coastal Systems Center | Bay | *Spartina — Scirpus* | 0.2 | 1977 | Sprigs |
| L. L. Marine | Escambia | *Spartina* | 0.02 | 1979 | *Spartina* sprigs |
| Florida Department of Transportation — Bayshore Park | Escambia | *Spartina — Scirpus)* | 0.2 | 1977 | Sprigs |

A major loss of coastal wetlands has been through filling activities to dispose of dredged material. The Federal Water Pollution Control Act (passed by Congress in 1972, amended in 1977, and now called the Clean Water Act) was predicated on the philosophy that no one has a right to pollute out Nation's waters. Section 404 of the Act gives the U.S. Army Corps of Engineers the permitting authority for discharge of dredged or fill material into wetlands and waters of the U.S. The 404 program has two paramount goals: (1) to protect and enhance the Nation's wetlands and other critical and sensitive aquatic resources and (2) to restore and maintain the chemical, physical, and biological integrity of the Nation's waters. The touchstone of the Section 404 permit process is compliance with environmental guidelines promulgated by EPA, which describe how to ascertain the potential environmental damage of dredged or fill material and the ecological value of the disposal site.

Filling of waters and wetlands without a Section 404 permit is an illegal activity and violators may be subject to civil and/or criminal fines, imprisonment, and/or injunctive relief including restoration of the area to its original condition. Restoration normally involves removal of fill material to preproject elevations and either allows the area to revegetate naturally or requires planting appropriate wetland species. Table 2 presents restoration projects of illegally filled areas for the portion of Florida covered by this chapter; the Mobile Corps of Engineers Office was not able to supply this information for Alabama and Mississippi. These restorations were supervised by the Corps of Engineers and are a potential source of useful data; however, none of these sites are currently being quantitatively monitored. A small effort in this regard could elucidate criteria necessary for successful establishment of functioning wetland systems on previously filled sites.

Federal dredging and filling activities in connection with a project specifically approved by Congress do not require a Section 404 permit, as long as environmental guidelines are considered in an environmental impact statement. Each year the Corps of Engineers dredges approximately 300 million yd³ of material from the 22,000 m of navigable waters of the continental U.S.[30] Clark[31] estimated that 34% of wetlands

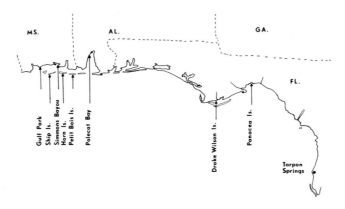

FIGURE 2.     Location of marsh creation studies on dredged spoil.

filied from 1955 to 1964 were filled to dispose of dredged material. A basis for the need for the restoration of such damaged ecosystems was substantially made by Cairns et al.[32] Today, the Corps of Engineers has recognized the detrimental effects of disposal on marshes and is seeking economically feasible disposal alternatives. One of the most viable alternatives to marsh disposal of spoil is the use of dredged material to create new marshes by planting spoil placed at intertidal elevations with salt marsh species. It is hoped that such efforts will rehabilitate and enhance the productivity of estuarine areas. Most of the pioneering work in this field has been by Woodhouse et al.[33,34] on the Atlantic coast.

## II. SALT MARSH CREATION

### A. Florida

Two attempts have been made in northwestern Florida to create salt marshes through planting of dredge spoil. Locations of the Dickerson Bay (Panacea) and Apalachicola Bay (Apalachicola) sites are shown in Figure 2. At Dickerson Bay, a 2-acre spoil island, Panacea Island, was created in October 1975 (at the edge of the dredged channel) from coarse-grained sandy dredged sediments. The intertidal area was planted in February 1976 with *J. roemerianus, D. spicata,* and *S. alterniflora* culms. The supratidal portion of the island was planted with *Uniola paniculata* and *Ammophila breviligulata* culms, and *Panicum amarum* seed in February 1976.[35,36]

Drake Wilson Island was created in Apalachicola Bay during the spring of 1976. A 7-acre intertidal area was enclosed by a containment dike which held 2 to 3 ft of fine-grained sandy sediments. A weir in the dike permitted tidal exchange and the area was planted in July 1976 with *S. alterniflora* culms in plots at 0.3, 0.6, 0.9, 1.8 and 2.7 m centers. *S. patens* was planted at elevations 0.9 to 1.8 m above mean sea level (MSL) at this site in July 1976.[36,37]

### B. Alabama

Transplanting of marsh grasses on dredged spoil was attempted in Polecat Bay (Figure 2), to determine the effect of marsh grasses on dewatering of dredged material.[38] Transplant materials included 2850 individual culms each of *S. alterniflora, S. cynosuroides, Phragmites communis,* and *Panicum repens.* These were planted from June 17, 1976, through July 8, 1976. Transplanting of all species was 100% unsuccessful in this study.

Dune areas at Dauphin Island have been successfully planted with *P. amarum* by Grant Mattox, U.S. Soil Conservation Service, and the Mobile County Soil and Water Conservation District.

## C. Mississippi

Eleuterius[39] and Eleuterius and Caldwell[40] have transplanted marsh and dune species on the Mississippi coast. Vascular plant species were planted on barren spoil areas at Horn Island Pass, Simmons Bayou, Gulf Park Estates Beach, Ocean Springs East Beach, and Ship, Horn, and Petit Bois Islands (Figure 2). Eleuterius chose ten species to study in Mississippi, all capable in his opinion of forming dense stands of vegetation depending on the site and other environmental conditions. Species used were *S. alterniflora, S. patens, S. cynosuroides, D. spicata, P. repens, P. amarum* var. *amarulum, P. communis, U. paniculata, Ipomoea stolonifera,* and *J. roemerianus.* Sods, sprigs, and cuttings rooted in peat pellets were used in Eleuterius' studies. *Panicum amarum* and *U. paniculata* were transplanted at dune sites, *P. repens* was transplanted at dune and marsh sites, and all other species were planted at marsh sites. Plantings throughout the year allowed an assessment of the optimum season for transplanting.

## III. PLANTED SPECIES

The above-referenced marsh creation studies undertaken in northwest Florida, Alabama, and Mississippi have yielded valuable information which will be useful in further transplanting attempts. The discussion below summarizes, by species, the findings of these studies. Eleuterius[40] presented no data for plantings of *I. stolonifera,* so although it was planted, no discussion of that species will be given herein.

### A. *Spartina alterniflora*

Throughout the intertidal portion of Panacea Island, 810, 5-in. long rhizomes were planted in plots. Approximately 50% of the plants were missing 1 month after planting and only a few clumps, scattered high in the intertidal plane, were present 17 months after planting. Most plants were washed out during the third month after planting by strong tidal currents and/or abrasion by coarse, shifting sands.[35]

*S. alterniflora* transplants survived well at Drake Wilson and achieved full cover in 0.3-, 0.6-, and 0.9-m spaced plants within 15 months after planting. Plants spaced at 1.8- and 2.7-m centers grew poorly (Figure 3). This may have been due to the proximity of the 1.8- and 2.7-m plots to the weir.[36]

Eleuterius[39] observed that *S. alterniflora* did not survive well when planted in peat pellets. Greatest survival of this species in Mississippi occurred in November (65%) and December (65%). No success was obtained with transplants made during April, June, September, and October. *S. alterniflora* was consistently the earliest colonizer of dredge spoil in Mississippi. Natural colonization ocurred from seed at a consistently well-defined position high in the intertidal plane. Plant colonies and immature stands spread centrifugally at first, then predominantly downward on the tidal plane from the point of establishment. Four or more years may be required before *S. alterniflora* reaches the lowest limit of vertical distribution. From a single transplant in Davis and Biloxi Bays, Miss.[40] two tenths of a hectare was completely covered by *S. alterniflora* in 2 years.

### B. *Spartina patens*

Bare root culms were obtained from nearby St. Vincents Island and planted on Drake Wilson Island in 9 by 9 m plots at spacings of 0.3, 0.9, 1.8, and 2.7 m. Rate of survival was correlated with spacing and elevation (Table 3). Plants spaced at 1.8 and 2.7 ft centers grew equally well and achieved taller size than those at 0.3 or 0.9 m spacings (Figure 4). Mean number of leaves per plant increased linearly with 0.3-, 0.9-, and 1.8-m spacing; mean number of leaves was nearly the same in 1.8- and 2.7-m plots. Xeric, nutrient-poor sediments can be rapidly stabilized through planting of this species at 1.8 and 2.7 m spacings at elevations ranging from 1.0 to 1.4 m above MSL.[37]

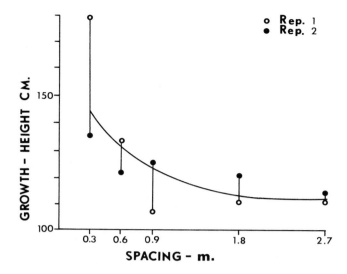

FIGURE 3.    Mean growth in height of new culms of *Spartina alterniflora* in all plots of both replicates on Drake Wilson Island from December 1976 to September 1977. Curve fitted by eye. (From Kruczynski, W. L. and Huffman, R. T., *Proc. 5th Annu. Conf. Restoration Coastal Vegetat. in Florida,* Tampa, Fla., 1978, 99. With permission.)

Eleuterius[39] found that cuttings of *S. patens* rooted very well in peat pellets, especially those planted in the greenhouse for rooting during the winter. Peat pellets showed higher overall survival (62%) than sprigs (44%). Greatest survival of this species occurred when planted during autumn and winter (October, 71%; February, 100%). Survival was best at Gulf Park Marsh (80%) and lowest at Simmons Bayou Marsh (27%).

## C. *Spartina cynosuroides*

Eleuterius[39] observed that this species did not root well in peat pellets and that best results were obtained through planting of culms. Highest survival (100%) was obtained for January (1974) transplants. However, no transplants planted in January 1973 survived. Survival at Gulf Park Marsh was 64% compared to 18% at Simmons Bayou Marsh. This species is slow growing and did not form closed stands 8 years after planting at Simmons Bayou.[41]

## D. *Juncus roemerianus*

This species exists in two forms, with either perfect flowers or pistillate flowers.[42] The distribution of the two flowering forms may be different in a natural marsh.[13]

Plantings at Panacea Island failed to become established and all transplants were missing 17 months after planting.[35]

The peat pellet method of rooting does not work well for *Juncus.*[39] Overall, sprigged plants survived better after planting (27%) than those rooted in peat pellets (2%). There was no apparent beneficial effect through treating cuttings with Hormodin rooting solution prior to planting. Eleuterius attempted to determine if there were differences in rooting and growth rates of the two flowering forms; however, all experimental cuttings kept in the laboratory were dead 4 weeks after planting in peat pots. Transplants of rhizomes in study plots demonstrated that both flowering forms had nearly the same overall survival rate, 17% compared to 16%. Transplants of the uni-

## Table 3
## MULTIPLE AND LINEAR REGRESSION COEFFICIENTS FOR ELEVATION AND SPACING WITH GROWTH VARIABLES OF *SPARTINA PATENS* PLANTED AT 0.3, 0.9, 1.8 m AT DRAKE WILSON ISLAND, FLA.

| Dependent variable | Time period | Independent variables | | |
|---|---|---|---|---|
| | | Spacing | Elevation | Both |
| Height increase | 12/76—9/77 | 0.64[a] | 0.64[a] | 0.80[b] |
| Leaves increase number | 12/76—9/77 | 0.96[b] | 0.27 | 0.96[b] |
| Clump width | 12/76—9/77 | 0.94[b] | 0.32 | 0.94[b] |
| Percent survival | 12/76—9/77 | 0.58[a] | 0.34 | 0.61[b] |
| Shoots number | 12/76—4/77 | 0.62[b] | −0.03 | 0.62[b] |
| Distance to furthest shoot | 12/76—4/77 | 0.70[b] | 0.48 | 0.81[b] |
| Height increase, tallest shoot | 12/76—4/77 | 0.41 | 0.16 | 0.47 |

[a] Significantly different from zero, $P<0.01$
[b] Significantly different from zero, $P<0.001$.

From Kruczynski and W. L., Huffman, R. T., *Proc. 5th Annu. Conf. Restoration Coastal Vegetation in Florida,* Tampa, Fla. 1978, 99. With permission.

sexual and bisexual flowering forms had 100 and 90% survival, respectively, when transplanted in January and February in Mississippi. Transplants of both sprigs and peat pellets of this species did not survive at Simmons Bayou Marsh.[39] Apparently, at a different site at Simmons Bayou Marsh, Eleuterius and Caldwell[40] reported that culms of *J. roemerianus* planted at 4-ft centers formed a closed stand in 5 years.

Mass planting of sods cut from a *Juncus* marsh showed excellent survival rates.[39] Eleuterius and Caldwell[40] observed that *Juncus* is very slow to naturally colonize spoil islands. It may take 10 years before this species begins to show up, often in the *D. spicata* zone. Frequently, *J. roemerianus* becomes established at the upper edge of the *S. alterniflora* zone, eventually crowding out that latter species. *Juncus* grows very slowly compared to *S. alterniflora* and *D. spicata,* but once it becomes established in an area, no other species seems to be able to displace it. Eleuterius and Caldwell[40] estimate that from 6 to 15 years may be required before *J. roemerianus* begins to colonize barren or vegetated spoil and from 10 to 20 years or longer before closed stands are formed.

Coultas[43] has transplanted this species in sand in a greenhouse. He found that transplants did not survive when planted in August and that treating roots with Rootone® before planting seemed to decrease survival.

### E. *Distichlis spicata*

*Distichlis* is best transplanted by rooting in peat pellets.[39] The species is dioecious and male plants showed a slightly greater overall survival (41%) than female plants (35%). Survival of this species, as with other marsh species, was lower at Simmons Bayou (male 28%, female 21%) compared to Gulf Park Marsh (male 53%, female 48%). Transplants made during February showed highest survival of both sexes. Two plots on Ship Island planted on 0.9 m centers in April 1974 showed 56 and 89% survival after 1 year.

*D. spicata* will rapidly colonize barren spoil islands and generally replaces *S. alterniflora* as that species grows down the tidal plane.[40]

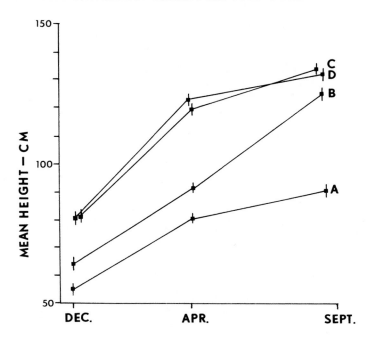

FIGURE 4. Mean height of tallest leaves of *Spartina patens* plants in plots A to D from three sample dates on Drake Wilson Island. Mean $+/-$ 2 $S_x$ indicated. Plant spacing: A — 0.3 m, B — 0.9 m, C — 1.8 m, D — 2.7 m. (From Kruczynski, W. L. and Huffman, R. T., *Proc. 5th Annu. Conf. Restoration Coastal Vegetation in Florida,* Tampa Fla., 1978, 99. With permission.)

## F. *Phragmites communis*

*Phragmites* was difficult to root in peat pellets and treatment with growth hormone before transplanting was not successful. Survival of sprigs was 29% at Gulf Park Marsh and 2% at Simmons Bayou Marsh. Survival was best for transplants in February, but seasonal results were inconsistent and generally less than 50%.[39] This species was found to be slow growing when transplanted by Eleuterius and Gill[41] and did not form a closed stand during 8 years of observation.

## G. *Panicum repens*

This species is best transplanted by rooting in peat pellets (74%) but sprigging also had high survival rates (65%). Best survival in Mississippi was obtained at dune sites (82%) as opposed to marsh sites (56%). Highest survival of transplants was in January (98%), February (94%), and March (92%). Overall survival at Gulf Park Marsh was 91%, while 42% of transplants survived at Simmons Bayou Marsh.

Mass planting of *P. repens* at Horn Island in August 1973 at dune sites (0 to 15 ft MSL) resulted in 91 and 80% survival in two plots after 1 year. A similar planting on Ship Island in April 1974 resulted in 63% survival of sprigs after 1 year.

## H. *Panicum amarum*

*P. amarum* was seeded in rows on the supratidal portion of Panacea Island and formed lush stands after 1 year. Standing crop biomass was increased with application of fertilizer (Figure 5).

*P. amarum* rooted easily in peat pellets and had consistently high survival at dune sites (62 to 90%) throughout the year in Mississippi except in March (28%). Highest

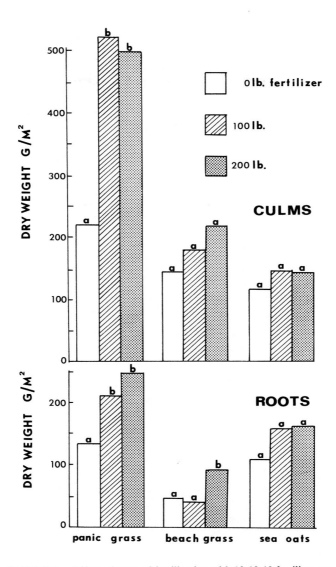

FIGURE 5. Effect of rates of fertilization with 10-10-10 fertilizer on above and below ground standing crop biomass of panic grass, beachgrass, and sea oats on Panacea Island. Like letters indicate no significant differences among treatments (ANOVA, P> 0.05); unlike letters indicate significant differences among treatments. (From Coultas, C. L., Breitenbeck, G. A., Kruczynski, W. L., and Subrahmanyum, C. B., *J. Soil Water Conserv.*, 33, 183, 1978. With permission.)

survival occurred with transplants made in January. Transplants rooted in peat pellets had a higher rate of survival (74%) than sprigs (53%). Survival of transplants was higher at Gulf Park (74%) compared to Simmons Bayou (59%).

A mass planting on 3-ft centers on Horn Island in 1975 had more than 50% dead after 6 months and poor survival (19 to 23%) after 1 year. Predation by rabbits was responsible for reducing growth and spread of these transplants. Sprigs planted at 3-ft centers in April 1974 on Ship Island formed a closed stand 3 months after transplanting. Survival there was 90 to 94% in replicate plots.[39]

Planting of *P. amarum* seed over approximately 20 acres of sand dunes at Dauphin

### Table 4
### EFFECTS OF FERTILIZATION ON THE
### HEIGHT AND SPREAD OF PANIC GRASS,
### BEACHGRASS, AND SEA OATS

| Plant | Fertilization rate | | |
|---|---|---|---|
| | 0 | 112 kg/ha | 224 kg/ha |
| *Panicum amarum* | | | |
| Height — cm | 69.6 a | 94.2 b | 93.5 b |
| Spread — cm | 9.0 a | 11.1 a | 9.0 a |
| *Ammophila breviligulata* | | | |
| Height — cm | 93.8 a | 95.1 a | 90.7 a |
| Spread — cm | 36.0 a | 34.4 a | 36.1 a |
| *Uniola paniculata* | | | |
| Height — cm | 98.2 a | 110.9 b | 118.4 b |
| Spread — cm | 42.4 a | 53.6 a | 56.7 a |

*Note:* Unlike letters (a,b) indicate significant difference, $p <$ 0.05, within each species.

From Coultas, C. L., Breitenbeck, G. A., Kruczynski, W. L., and Subrahmanyum, C. B., *J. Soil Water Conserv.*, 33, 183, 1978. With permission.

Island was successful. Seeding rate was 20 lb/acre and was raked in with a spike-tooth harrow. Fertilizer was applied yearly by air for 3 years at the rate of 448 kg/ha of 13-13-13 (N-P-K). Fertilizer application has increased growth of this species, but this was not quantitatively measured.[44]

### I. *Uniola paniculata*

Survival of culms planted at Panacea Island was 44 to 88% by rows. Fertilizer treatment had no significant effect on survival; however, maximum heights of sea oats were greater in fertilized treatments (Table 4).

Sea oats proved difficult to root in peat pellets. Eleuterius[39] observed that hormone treatment seemed to increase root length after planting of this species, but no statistical analysis was performed. Overall, survival at dune sites of culms was 38%. No peat pellet transplants survived, while 48% of sprigs did. Transplants made during March demonstrated highest survival (85%). In contrast to most species planted in Mississippi by Eleuterius, sea oats survived best at Simmons Bayou (43%) compared to Gulf Park (22%).

### J. *Ammophila breviligulata*

American beachgrass is widely used for dune stabilization on the Atlantic coast. It does not occur on the Gulf coast but was successfully transplanted by Coultas et al.[35] at Panacea Island. Survival of culms varied between 54 and 83% by rows. Height and spread was not significantly greater in fertilized rows (Table 4).

## IV. DISCUSSION

### A. Handling Cuttings

Transporting large clumps of marsh grasses over long distances causes significant logistic problems. Thus, culms are commonly used in planting studies. Transplant material was gathered from adjacent marshes in the programs at Panacea, Apalachicola, Polecat Bay, Alabama, and Mississippi. Culms were transplanted as soon as possible after being removed from donor sites to minimize physiological shock. Eleuterius[39] however, found that *Panicum* culms and cuttings can be stored for several months in plastic bags at 40°F and still produce viable, rooted transplants.

Eleuterius[39] achieved excellent results growing cuttings of *Panicum amarum, P. repens, Spartina patens,* and *Distichlis spicata* in peat pellets. This method had little value for *J. roemerianus* and was unsuccessful for *Uniola paniculata, S. cynosuroides, Phragmites communis,* and *S. alterniflora.* Transplanting cuttings rooted in peat pellets minimizes transplantation shock and trays of peat pellets are easy to transport. Eleuterius recommends growing shoots for 3 to 4 months in peat pellets before transplanting since this produces vigorous transplants which have greater survival and growth rates than newly planted cuttings or culms. If kept longer than 3 or 4 months, cuttings become cumbersome to handle.

Transplanted material rooted best in peat pellets when maintained under constant light conditions at 72°F. A "hardening off" period, gained through exposure of rooted cuttings to outdoor conditions before transplanting, increased survival. *Panicum repens* and *P. amarum* may be pruned before transplanting; pruning reduces the amount of water needed by the plant and provides a smaller, more vigorous, easier to handle transplant.

## B. Hormone Treatment

Several authors have tried to increase the success of transplanting through treatment of transplants with growth hormones. Preliminary experiments by Eleuterius[39] indicate that treatment with 10% napthaleneacetic acid solution retarded root initiation in all species except *Uniola.* Root growth of *Uniola* increased after treatment. Coultas[45] treated rhizomes and roots of *J. roemerianus* with Rootone® and observed no statistical difference between growth and survival of treated and untreated transplants. One plot of each species planted at Polecat Bay, Ala. was treated with Up Start® (however, no plants survived in treated or untreated plots in that study.)

## C. Fertilization

Replicate rows of the supratidal portion of Panacea Island were fertilized with 0, 112, or 224 kg/ha of 10-10-10 (N, P, K) fertilizer in April and July 1976. Fertilizer had little effect on survival of *P. amarum.* Height and weight of *P. amarum* roots and culms responded to 112 kg/ha treatment (Table 4 and Figure 5). The 224 kg/ha rate did not increase the response over the 112 kg treatment. Fertilizer did not significantly increase height or spread of *Ammophila* (Table 4). Root and rhizome growth was greatest following 224 kg/ha application. Height of *Uniola* significantly increased following application of 112 and 224 kg/ha. Although spread and weight of culms and weight of rhizomes did not increase significantly with fertilization, biomass yields rose (Table 4 and Figure 5).

Eleuterius[39] found that dune sites at Simmons Bayou and Gulf Park Estates Beach had low levels of nutrients, primarily phosphorus. Two slow release nutrient sources (Mag-Amp®, 6-40-40 and Osmocote®, 40-6-4) were applied at the rate of 112 kg/ha on two different occasions. No observable difference in growth was noted. Nutrients were also applied to Ocean Springs East Beach and Horn Island plots, but growth was extremely low compared to growth of transplants on unfertilized, but recently deposited, spoil at Ship Island. Eleuterius concluded that hypersalinity of soil water adversely affected plant growth at Ocean Springs East Beach, while low soil moisture and rabbits were major factors affecting growth at Horn Island.

## D. Planting Methods

All plantings discussed for this geographic area were hand planted. Eleuterius[39] stated that mechanized planting of rooted cuttings set in peat pellets could result in

### Table 5
### MAN HOURS REQUIRED TO TRANSPLANT ONE ACRE OF SPOIL

| Species | Digging | Sorting/ washing | Prepare peat pellets | Maintaining 6 weeks | Planting | Total |
|---------|---------|------------------|----------------------|---------------------|----------|-------|
| *Panicum repens*[a] | 32 | 24 | 40 | 6 | 50 | 152 |
| *Panicum amarum*[a] | 16 | 50 | 20 | 6 | 50 | 142 |
| *Spartina alterniflora*[a] | 28 | 6 | — | — | 88 | 122 |
| *Spartina cynosuroides*[b] | 60 | 32 | — | — | 46 | 138 |
| *Panicum repens*[b] | 6 | 18 | | | 42 | 66 |
| *Phragmites communis*[b] | 216 | — | — | — | 40 | 256 |

[a]  From Eleuterius, 1974.[39] Data based on preparation of 5000 transplants in peat pellets, with field emplacement at 0.9 m centers (11,818 transplants per hectare).

[b]  From Stout, 1977[38] Data based on collecting and planting 1550 transplants with field emplacement at 0.3, 0.9, 1.8, 2.7 m centers.

rapid, complete cover of an area. Two studies utilized a Riverine Utility Craft (RUC) to traverse soft, fluid-like dredged material. Kruczynski et al.[39] utilized the RUC to tow a plywood sled that served as a platform to plant *S. alterniflora* sprigs at Drake Wilson Island. The surface crust and existing vegetation of the spoil site at Polecat Bay were turned under by repeated passes of a RUC before planting.[38] The use of a RUC is recommended where environmental conditions assure that closeness and exactness of plant spacing for optimum growth in a specified period of time is not a major factor for success.

Manual transplanting of marsh species directly to test plots required digging of transplants from an existing marsh, separation of transplant material into individual rhizome pieces, and replanting in prepared plots at the test sites. Man-hours required for each of the four species planted at Polecat Bay are given in Table 5. *Panicum repens* proved to be the least costly of the species used in this study since it was the easiest to dig and plant (66 man-hours to dig, sort, and plant 1550 transplants). Eleuterius[39] reported that 152 and 142 man hours were necessary to transplant 0.4 ha of *P. repens* and *P. amarum*, respectively (Table 5). This estimate is for planting 4785 prepared peat pellets per acre and does not include costs of materials, transportation time, nor preparation of necessary propagating facilities for growing cuttings in peat pellets.

### E. Factors Affecting Success of Transplants

Factors affecting success of transplant survival listed in decreasing order of importance, according to Eleuterius[39] are erosion by either wind, water, or both; available soil water; soil water salinity; and soil nutrients. If these four criteria are favorable, transplanting should proceed with little difficulty and lead to predictable success.

Erosion was the major factor in lack of success among intertidal plots at Panacea Island. Proximity to the tidal weir at Drake Wilson Island was thought to be the main reason for poor success of 1.8- and 2.7-m spaced transplants of *S. alterniflora*.[36] Also, approximately 1000 sod and sprig transplants were lost by Eleuterius because they were planted upon spoil and beaches exposed to high energy waves during winter months. Excessive erosion, i.e., elevational reductions of about 0.6 m or more per year, or 30 m of shoreline loss, are good indicators of sites to be avoided for transplantation studies. Eleuterius recommends that elevation and topographic information be obtained for at least 1 year prior to any transplanting operation. On newly deposited spoil, 2 years are necessary since at least 1 year is necessary for normal settling of sediments. Eleuterius and Gill[41] have successfully used hay bales to form a barrier against waves and currents. Use of hay bale allows tidal flooding without erosion and is recommended around spoil deposited in shallow waters.

Eleuterius and Caldwell[40] observed that natural colonization on Mississippi spoil islands occurs high on the intertidal plane. Plant species grow downward on the tidal plane to reach the zone of their most vigorous growth and density. This is true for grasses, shrubs, and trees. The exact cause of this phenomenon is unknown, but it indicates that higher regions of low profile spoil are more favorable to seedling establishment and subsequent growth than lower areas. This phenomenon was consistent for 12 years of observation, even though lower intertidal areas are more constant in environmental conditions and higher regions are subject to greater fluctuations in soil water, salinity, and nutrient levels. These findings have a direct bearing on transplanting work, and Eleuterius and Caldwell[40] and Eleuterius and Gill[41] suggest that it is ecologically and economically advantageous to transplant all species higher on the intertidal plane than they are found in adjacent natural marshes and let the plants adjust themselves downward until they reach locations and positions for optimum growth.

Transplants at Gulf Park Estates Beach, had a greater survival rate (63%) than those at Simmons Bayou (28%). This difference was thought to be due to higher salinity of soil water at Simmons Bayou. Eleuterius measured surface soil salinity as high as 144°/$_{oo}$ at Simmons Bayou. Fluctuation of soil moisture was also thought to be a factor in the relatively poor success at Simmons Bayou.

Stout[38] theorized that lack of success at Polecat Bay was due to time of year and air temperature during planting. Temperatures were in the 35°C range and rainfall was minimal during and after planting. However, S. alterniflora and S. patens showed excellent survival when transplanted in July 1976 at Drake Wilson Island. Transplanting success at Drake Wilson Island is also not consistent with seasonal success observed in Mississippi. Eleuterius[39] found that late autumn through early spring was the best time to transplant marsh species. Panicum repens showed best survival in January to March, S. cynosuroides in January, P. communis in February, S. alterniflora in November to December. Coultas[43] stated that J. roemerianus did not survive when transplanted into a greenhouse in August; it did survive when transplanted during all other months.

## F. Natural Colonization

Stout[38] found that natural vegetation flourished at the North Blakely spoil site in Alabama and produced a dense surface cover and network of roots and rhizomes. Five species appeared as the dominant vegetation at control plots: Panicum dichotomiflorum, Pluchia, Aster, Amaranthus cannebinus, and Phragmites communis. Kruczynski et al.[13] found that 42 species invaded Drake Wilson Island during the period of study. However, they recommended against reliance upon natural invasion to initially stabilize dredged material sites since development of sufficient ground cover is often happenstance.

Selection of plant species that provide ground cover in a short period of time, stabilize substrate, and become a predominant primary producer for the estuarine food chain should be the prime concern in any habitat development plant for a dredged material disposal site. Eleuterius[39] observed that, of the numerous species which are naturally found on spoil banks, only a few have growth forms suitable for rapid spread or spoil binding, attributes worthy for use in marsh creation or dune stabilization. Of the species he studied, overall survival at all sites in descending order, was Panicum repens (70%), P. amarum (66%), Spartina patens (53%), D. spicata male (41%), S. cynosuroides (39%), U. paniculata (38%), D. spicata female (35%), S. alterniflora (27%), Phragmites communis (26%), J. roemerianus bisexual (17%), and J. roemerianus female (16%). Eleuterius and Gill[41] found that in randomly planted plots, S. patens, S. alterniflora, Panicum repens, and D. spicata formed closed stands within a year. Spartina cyosuroides and Phragmites communis did not grow appreciably during 3 years of observation and did not form closed stands 8 years after planting.

## G. Hardwood Transplants

Although all of the known published work for this geographic area has been on transplanting herbaceous species, one attempt is known in which wetland hardwood tree saplings have been transplanted (Table 2). The Corps of Engineers directed L. Weisman to restore a 1.5-acre portion of a hardwood swamp which was destroyed by the construction of a road. This site is adjacent to the Crystal River in Florida. Bare-root transplants of red maple were planted on 10-ft centers in 1981. No data are being gathered on growth and survival of transplants. However, the site is occasionally visited by Corps and EPA personnel so that hardwood trees which are observed to do best under these conditions can be recommended for future restoration projects in hardwood wetlands.

## REFERENCES

1. **Humm, H. J., III**, The biological environment, A. Salt marshes, in *A Summary of Knowledge of the Eastern Gulf of Mexico*, Jones, J. I., Ring, R. E., Rinkel, M. O., and Smith, R. E., Eds., State University System of Florida, Institute of Oceanography, St. Petersburg, 1973.
2. **Eleuterius, L. N.**, The marshes of Mississippi, *Castanea*, 37, 153, 1972.
3. **Eleuterius, L. N. and McDaniel, S.**, The salt marsh flora of Mississippi, *Castanea*, 43, 86, 1978.
4. **Lindall, W. N. and Saloman, C. H.**, Alteration and destruction of estuaries affecting fishery resources of the Gulf of Mexico, *Mar. Fish. Rev.*, 39, 1, 1977.
5. **Thayer, G. W. and Ustach, J. F.**, Gulf of Mexico Wetlands; Value, state of knowledge and research needs, *Proc. Symp. on Environ. Res. Needs in the Gulf of Mexico (GOMEX)*, Vol. IIB, U.S. Department of Commerce, Atlantic Oceanographic and Meteorological Laboratories, Miami, 1981, 1.
6. **Eleuterius, L. N.**, The distribution of *Juncus roemerianus* in salt marshes of North America, *Chesapeake Sci.*, 17, 289, 1976.
7. **Kurz, H. and Wagner, K.**, Tidal marshes of the Gulf and Atlantic coasts of northern Florida and Charleston, South Carolina, Florida State University, Tallahassee, Studies No. 24, 1957.
8. **Adams, D. A.**, Factors influencing vascular plant zonation in North Carolina salt marshes, *Ecology*, 44, 445, 1963.
9. **Charbreck, R. H.**, Marsh Zones and Vegetative Types in Louisiana Coastal Marshes, Ph.D. dissertation, Louisiana State University, Baton Rouge, 1970.
10. **Kurz, H.**, Vegetation of northwest Florida tidal marshes, Final Rep., Office of Naval Research, Contract, NR 163-217, Arlington, Va., 1953.
11. **Seneca, E. D., Broome, S. W., Woodhouse, W. W., Cammen L. M., and Lyon, J. T., III**, Establishing *Spartina alterniflora* marsh in North Carolina, *Env. Cons.*, 3(3), 185, 1976.
12. **Turner, R. E.**, Geographic variations in salt marsh macrophyte production: a review, *Contrib. Mar. Sci.*, 20, 48, 1976.
13. **Kruczynski, W. L., Subrahmanyum, C. B., and Drake, S. H.**, Studies on the plant community of a North Florida salt marsh, I. Primary production, *Bull. Mar. Sci.*, 28, 316, 1978.
14. **Mendelsson, I. A., McKee, K. L., and Patrick, W. H., Jr.**, Oxygen deficiency in *Spartina alterniflora* roots: metabolic adaptation to anoxia, *Science*, 214, 439, 1981.
15. **Gabriel, B. C. and de la Cruz, A. A.**, Species composition, standing stock, and net primary productivity of a salt marsh community in Mississippi *Chesapeake Sci.*, 15, 72, 1974.
16. **Gosselink, J. G., Hopkinson, C. S., Jr., and Parrondo, R. T.**, Common marsh plant species of the Gulf coast area, in Productivity, Vol. 1, U.S. Army Corps of Engineers, Waterways Exp. Stn., Tech. Rep. D-77-44, Vicksburg, Miss., 1977.
17. **Wiegert, R. G. and Evans, F. C.**, Primary production and the disappearance of dead vegetation in an old field in southeastern Michigan, *Ecology*, 45, 49, 1964.
18. **Subrahmanyum, C. B., Kruczynski, W. L., and Drake, S. H.**, Studies on the animal communities in two North Florida salt marshes, II. Macroinvertebrate communities, *Bull. Mar. Sci.*, 26, 172, 1976.
19. **Kraeuter, J. N. and Wolf, P. L.**, The relationship of marsh macroinvertebrates to salt marsh plants, in *Ecology of Halophytes*, Riemold, R. J. and Queen, W. H., Eds. Academic Press, New York, 1974.

20. **Subrahmanyum, C. B. and Drake, S. H.**, Studies on the animal communities in two North Florida salt marshes, I. Fish communities, *Bull. Mar. Sci.*, 25, 445, 1975.

21. **Teal, J. M.**, Energy flow in a salt marsh ecosystem of Georgia, *Ecology*, 43, 614, 1962.

22. **Odum, E. P. and de la Cruz, A. A.**, Particulate detritus in a Georgia salt marsh — estuarine ecosystem, in *Estuaries*, Lauff, G. H., Ed., Am. Assoc. Adv. Sci. Publ. 83, Washington, D.C. 1967.

23. **Heald, E. J.**, The Production of Organic Detritus in a South Florida Estuary, Ph.D. dissertation, University of Miami, 1969.

24. **Herke, W. H.**, Use of Natural and Semi-impounded, Louisiana Tidal Marshes and Nurseries for Fishes and Crustaceans, Ph.D dissertation, Lousiana State University, Baton Rouge, 1971.

25. **de la Cruz, A. A.**, Recent advances in our understanding of salt marsh ecology, in Proc. Gulf of Mexico Coastal Ecosystems, Workshop, Fare, P. L. and Peterson, R. D., Eds., U.S. Fish and Wildlife Service, FWS/OBS-80/80, 1980, 51.

26. **McHugh, J. L.**, Estuarine fisheries: are they doomed?, in *Estuarine Processes*, Vol. 1, Wiley, M., Ed., Academic Press, New York, 1976.

27. Council on Environmental Quality, Our Nation's Wetlands, An Interagency Task Force Report, U. S. Government Printing Office, Washington, D.C., 1978.

28. United States Environmental Protection Agency, Southeast Environmental Profiles 1977, U.S.E.P.A. Region IV, Atlanta, Ga., 1978.

29. **Shaw, S. P. and Fredine, C. G.**, Wetlands of the United States, their extent and their value to waterfowl and other wildlife, Circ. 69, U.S. Fish and Wildlife Service, U.S. Government Printing Office, Washington, D.C., 1971.

30. **Boyd M. B., Saucier, R. T., Keeley, J. W., Montgomery, R. L., Brown, E. O., Mathis, D. B., and Guice, C. J.**, Disposal of dredge spoil, U. S. Army Corps of Engineers, Waterways Exp. Stn. Tech. Rep. H-72-8, Vicksburg, Miss., 1972.

31. **Clark, J.**, *Fish and Man: Conflict in the Atlantic Estuaries*, American Littoral Society, Spec. Publ. No. 5, Highlands, N. J., 1967.

32. **Cairns, J., Jr., Dickson, K. L., and Herricks, E. E.**, Eds., *Proc. Int. Symp. Recovery Damaged Ecosystems*, Virginia Polytechnic Institute and Virginia State University, University of Virginia Charlottesville, Press, 1975.

33. **Woodhouse, W. W., Seneca, E. D., Jr., and Broome, S. W.**, Marsh building with dredge spoil in North Carolina, Bull. 445, Agric. Exp. Stn., North Carolina State University, Raleigh, July 1972.

34. **Woodhouse, W. W., Jr., Seneca, E. D., and Broome, S. W.**, Propagation and use of *Spartina alterniflora* for shoreline erosion abatement, TR 76-2, U. S. Army Corps of Engineers, Coastal Eng. Res. Center, Fort Belvoir, Va., August 1976.

35. **Coultas, C. L., Breitenbeck, G. A., Kruczynski, W. L., and Subrahmanyum, C. B.**, Vegetative stabilization of dredge spoil in North Florida, *J. Soil Water Conserv.*, 33, 183, 1978.

36. **Kruczynski, W. L. and R. T. Huffman**, Use of selected marsh and dune plants in stabilizing dredged material at Panacea and Apalachicola, Fla., *Proc. 5th Annu. Conf. Restoration of Coastal Vegetation in Florida*, Tampa, Fla., 1978, 99.

37. **Kruczynski, W. L., Huffman, R. T., and Vincent, M. K.**, Habitat development field investigations, Apalachicola Bay marsh development site, Apalachicola, Fla., U. S. Army Corps of Engineers, Waterways Exp. Stn., Tech. Rep. D-78-32, Vicksburg, Miss., 1978.

38. **Stout, J. P.**, Evaluation of plants as medium for dredged material dewatering, unpublished rep. to the U. S. Army Corps of Engineers, Mobile District, Mobile, Ala., Contract No. DAC-WO1-76-0170, 1977.

39. **Eleuterius, L. N.**, A study of plant establishment on spoil areas in Mississippi Sound at adjacent waters, U. S. Army Corps of Engineers Field Rep., DA (NO1-72-C-0001), 1974.

40. **Eleuterius, L. N. and Caldwell, J. D.**, Colonizing patterns of tidal marsh plants and vegetational succession on dredge spoil, Proc. 8th Annu. Conf. Wetlands Restor. Creation, in press.

41. **Eleuterius, L. N. and Gill, J. I., Jr.**, Long-term observations on seagrass beds and salt marsh established from transplants, Proc. 8th Annu. Conf. Wetlands Restor. Creation, in press.

42. **Eleuterius, L. N.**, Flower morphology and plant types within *Juncus roemerianus*, *Bull. Mar. Sci.*, 24, 493, 1974.

43. **Coultas, C. L.**, personal communication, 1977.

44. **Mattox, G.**, personal communication, 1981.

45. **Coultas, C. L.**, personal communication.

Chapter 4

SALT MARSHES OF THE WESTERN GULF OF MEXICO

James W. Webb, Jr.

TABLE OF CONTENTS

## I. DESCRIPTION OF PLANT COMMUNITY

Coastal marshes along the Gulf of Mexico in Louisiana and Texas can be divided into three major ecosystems. The Mississippi Deltaic plain ecosystem extends eastward from Vermilion Bay, La. to the Alabama-Mississippi state line.[1] The Chenier Plain Ecosystem extends from Vermilion Bay, La. to East Bay, Tex.[2] The Texas Barrier Island Ecosystem extends from East Bay to the U.S.-Mexico border.

This area is a complex mixture of open water, fresh to saline wetlands, and uplands. A number of classification systems have been used to divide the tidally influenced emergent vegetation into various habitat types. Gosselink et al.[2] divided the emergent wetlands of the Chenier plain into six categories: salt marsh, brackish marsh, intermediate marsh, fresh marsh, swamp forest, and impounded marsh. Two categories, salt marsh and brackish marsh, are tidally influenced and dominated by salt tolerant plants. Allan,[3] in his discussion of Gulf Coast marshland succession, stated that only two types of emergent coastal marsh communities occur, salt and fresh. To avoid confusion, he omitted the term "brackish". In the mapping conventions of the U.S. Fish and Wildlife Service, which are based on the work of Cowardin et al,[4] the term "brackish", is also omitted and only the terms estuarine and palustrine (fresh) are used to separate emergent type wetlands of the U.S. The term "brackish", which may refer to various salinity levels, is not utilized in this system. Since the U.S. Fish and Wildlife mapping conventions can be applied nationwide without regard to plant species present, this terminology seems the most applicable of the systems. The terminology in this chapter, however, follows that of each author cited.

The U.S. Department of Interior wetland classification system separates the estuarine emergent wetland habitats into a regularly flooded zone of *Spartina alterniflora* (smooth cordgrass) and an irregularly flooded zone of *S. patens* (marshhay cordgrass) and associated species.[4] These habitat types roughly correspond to the low and high marsh categories sometimes used on the Atlantic Coast.

Coastal marsh plants of the Gulf coast are generally the same dominant species as those found on the Atlantic coast of North America and occupy the same relative position with respect to water table and salinity levels.[3,5-8] *S. alterniflora* is the dominant plant in the intertidal zone. Generally, this species occurs from approximate mean high water (MHW) to mean low water (MLW). Ponding and poor drainage of tidal waters associated with spring tides and wind pushed tides may allow occurrence of this species at higher elevations. A narrow zone of bare soil may occur above *S. alterniflora*. At the next highest elevation, *Distichlis spicata* (saltgrass) occurs in a narrow zone.[3] *S. patens* normally dominates salt marshes at elevations just above *D. spicata*. Vast acreages of marsh dominated by *S. patens* occur throughout Texas and Louisiana. *Scirpus maritimus* (leafy threesquare) and *D. spicata* commonly occur with *Spartina patens*. Very wet areas or burned areas of the upper elevation salt marsh may be dominated by *Scirpus olneyi* (Olney bulrush). Many other species, such as *Borrichia frutescens* (sea oxeye daisy), *Monanthochloe littoralis* (saltflat grass), *Limonium carolinianum* (sea lavender), *Batis maritima* (saltwort), *Salicornia bigelovii* (annual glasswort), *S. virginica* (virginia glasswort), *Iva frutescens* (bigleaf sumpweed), *Scirpus olneyi,* and *S. maritimus,* occur in this infrequently flooded zone. Each of these species may achieve dominance in local areas. *B. maritima,* in particular, often dominates extensive areas of coastal flats.

The extensive salt marshes present on the upper Texas coast and in Louisiana change southward of Corpus Christi, Tex. into extensive wind-tidal flats.[9] Salt marsh with *S. alterniflora* present is essentially absent along the margins of the tidal flats of Laguna Madre and Baffin Bay due to erratic tide levels and hypersaline water. Subaqueous marine grassflats occur below the intertidal zone, adjacent to the tidal flats.[10] High marsh areas composed of *B. maritima, M. littoralis, Salicornia virginica, Lycium carolinianum* (wolfberry), *Borrichia frutescens,* and *Spartina patens* occur on many areas of the shoreline.

Stands of *Avicennia germinans* (black mangrove) occur occasionally in Louisiana and the upper Texas coast, but cold winters kill many of the plants periodically. Therefore, extensive stands of *A. germinans* are seldom seen above Corpus Christi. Large areas of this species are present around Corpus Christi Bay.

Along the Texas and Louisiana coast the irregularly flooded salt marsh zones may either change abruptly into or intergrade gradually into uplands or fresh water marshes. Historically, upland forests have not been extensive in the Chenier Plain.[11] In Texas, the term Gulf Prairies and Marshes is used to describe the coastal zone, which is dominated by grasses and sedges.[12] However, many upland areas are dominated by scrub/shrub species.

## II. PRODUCTIVITY AND HABITAT VALUES

The amount of emergent tidal vegetation within the Gulf of Mexico estuarine area makes it a significant habitat because of its many values. There are 6,075,299 acres of emergent tidal vegetation as opposed to 796,796 acres of submerged grass beds, and 7,890,611 acres of open estuarine water. Of the emergent vegetation areas in the Gulf of Mexico, 64% occurs in Louisiana while 19% occurs in Texas.[13] Louisiana alone, has 30% of the nations coastal wetlands, which includes freshwater wetlands.[14]

Despite the 370 mi (595 km) of Texas coastline and the 2498 mi of Gulf, bay, and estuary shorelines at the mean high tide line, Louisiana has much more tidal marsh than Texas.[15,16] However, there is conflicting information on the amount of marsh present. Griffitts[17] listed 3,381,500 acres in Louisiana while Lindall and Soloman[13] listed 3.9 million acres. In Texas, there is a large discrepancy in data. Griffitts[17] stated that there were 315,000 acres. Diener[15] tabulated 884,000 acres while Lindall and Soloman[13] list 1.1 million acres.

Coastal marshes of the western Gulf region have many values. Marshes provide a nursery and protective habitat for many finfish and shellfish, of which many are commercially and recreationally important. In the study by Peters et al.,[18] 16 wetland associated groups composed 59% of the U.S. commercial landings by weight. Most of the species utilize estuarine wetlands during young and larval stages of their life cycle. One half of salt water recreational fisheries, in terms of angler days per year, was directed toward wetland associated species.

Shorelines with emergent vegetation appear to provide excellent habitat as compared to unvegetated shorelines. For example, Mock[19] found that shrimp preferred an unaltered vegetated shoreline to a bulkheaded area in Clear Lake, Tex. Conner and Truesdale[20] reported that a low salinity mash area in the Galveston Bay system was prime habitat for the postlarva stage and juveniles of several marine crustaceans and fishes, including commercial shrimps, Gulf menhaden, Atlantic croakers, sand seatrout, and southern flounder. The harvest of penaied shrimp in Louisiana and other areas was positively correlated to the amount of estuarine marshland, but it was not correlated with the amount of estuarine waters in each locality.[21]

Avian utilization of coastal marshes is very high. Waterfowl utilize coastal marshes heavily. Approximately 250,000 geese were recorded on the Texas coast in 1939 to 1941.[22] About half of the wintering population of the central flyway has been reported in the coastal zone of Texas.[23] Wading and shore birds are common inhabitants of coastal marshes. On a dredge spoil area in Galveston Bay 40 out of 98 avian species sighted utilized the shoreline.[24] The salt marsh vegetation enhances utilization by shorebirds. For example, after creation of a salt marsh, 21 species of shorebirds utilized the salt marsh as compared to 18 on an adjacent natural shoreline.[25]

Amphibian and reptilian utilization of salt marshes has been poorly documented. However, many amphibians and reptiles habitually or occasionally invade brackish or salt water habitats.[26] The American alligator, an endangered species which is commercially important for its hides, utilizes fresh to brackish marshes of the coast. The diamondback terrapin is restricted in habitat to salt marshes.

The salt marsh environment provides habitat for many species of mammals, including commercially important furbearers such as the river otter, muskrat, nutria, and raccoon.[27] Marsh management techniques such as those developed for muskrats enhance economic returns from some furbearer species.[28,29] Small rodents of the salt marsh such as the hispid cotton rat and marsh rice rat, along with young nutria and muskrats are important food sources for carnivores and birds of prey. Marsh hawks and other birds of prey are common winter inhabitants of salt marshes of the western Gulf of Mexico.

Invertebrate utilization of salt marshes is important in the estuarine and associated wetland food chain. Drum feed on small crabs and shrimp.[30] Raccoons feed on fiddler crabs and mollusks.[31,32] Waterfowl feed on insects.[33] Mollusks, crustaceans, and other invertebrates are food items for the clapper rail, greater yellowlegs, waterfowl, fish, and marine animals.[32]

Detritus production by emergent vegetation could possibly be the most important value of salt marshes. Many authors have reported that emergent marsh plants may be one of the most important primary producers in estuaries.[34-37] Net primary production varies considerably by species and area. *Spartina patens* could be the greatest producer (4200 g/m²/year.[38] However, the occurrence of *S. patens* at higher elevations than *S. alterniflora* may prevent much of the plant material from entering the estuaries as detritus. Heinle and Flemer[39] reported that virtually no carbon was exchanged on a poorly flooded tidal marsh in Maryland. *S. alterniflora* may be the highest detritus producing species in the estuary because of its occurrence at elevations that are regularly flooded. Export of detritus from salt marshes in Calcasieu Lake in Louisiana was 7300 kg/ha/year as compared to 1100 kg/ha/year from fresh water marshes.[2]

Production values for *S. alterniflora* are often misleading because of failure to report elevations of collections in Louisiana and Texas. Webb et al.[40] reported that the biomass of *S. alterniflora* was four times greater at mid-tide range than at MHW. Differences in measurements for streamside and high marsh *S. alterniflora* have been reported in North Carolina.[41] Generally, production of *S. alterniflora* will be similar to the range of 500 to 2800 g/m²/year reported by Odum et al.,[42] but production of organic matter in salt marshes is increased by growth of organisms on surfaces of stems and leaves.

The esthetic values of salt marshes are often not appreciated by the general population. Historically, all marshes were considered wastelands that should be filled to increase agricultural and industrial production and to eliminate mosquitoes.[43] Recently, however, the unique habitats associated with salt marshes have made them a popular retreat for those seeking to birdwatch, explore, or simply enjoy the remoteness from noise, pollution, and people.

There are some indications that the high scenic quality of wetlands is not paralleled by high recreational use.[44] Cheek and Field[45] found fewer recreational activities and less activity overall associated with swamp and marsh environments, as opposed to other environments. They also found that nature study and food gathering activities were the most significant recreation activities. Nevertheless, recreation opportunities associated with food gathering activities are numerous. Waterfowl hunting alone, can be separated into various types of sporting activities, such as duck shooting over decoys, pass shooting ducks, jump shooting ducks from potholes and creeks, decoying or pass shooting geese, jump shooting marsh hens as they are flushed while walking or poling a boat through tidally flooded marsh grass, and jump shooting snipe. Rabbit hunting can be done with or without dogs. Fishing opportunities include rod and reel angling for flounder, trout, red fish, croaker, and other fish; gigging for flounder; use of trot lines for trout, redfish, and catfish; crabbing; and use of various types of seines and nets for fish, shrimp, and crabs. Oyster collecting often occurs in streams throughout salt marshes while clams may be collected on mud flats. In addition to food gathering activities, photography, hiking, marsh walks, canoeing, and boating are popular activities of salt marshes.

Salt marshes are valuable erosion control agents. They buffer inland areas from wave action since stems and leaves reduce current velocity. Emergent vegetation such as *S. alterniflora* protects shorelines from normal wave action while other plants may be important in slowing wave action during times of wind pushed spring tides, severe storms, or hurricanes. The marshes are floodwater reservoirs that reduce flooding in upland areas.[2] The reduction of current velocity by plants promotes sedimentation of organic and inorganic particles in estuarine waters. The root system tends to bind sediments and prevent resuspension of particles. Scouring of passes may be decreased by reduction in water currents as intertidal volume decreases.[46]

Salt marshes as well as entire estuaries may become valuable in tertiary sewage treatment in urbanized areas. One hectare of marsh-estuary is capable of performing $14,000 worth of tertiary treatment per year at a daily loading of nutrients equivalent to 8.8 kg of biological oxygen demand. This cost estimate is based on an artificial treatment cost of $1.6/kg biological oxygen demand.[2]

Salt marshes, in addition to sewage treatment, are valuable in purifying and buffering estuaries against large changes in upstream inputs of nutrients and wastes. Wetland sediments are able to reduce metals and organic toxins in flooding waters.[47]

## III. HABITAT LOSS AND MODIFICATION

Habitat loss and modifications to the habitat can be attributed to either natural or man-induced causes. Natural land loss occurs primarily by erosion and subsidence. In both cases, marsh loss may be accelerated by man's influence. Erosion is a considerable problem in many estuaries. Much of the erosion occurs on the northwestern shores of the bays due to the prevailing southeasterly winds much of the warm months of the year.[48] Bay and estuary shorelines in Texas were classified according to the amount of erosion occurring in 1968.[49] Erosion was critical on 57 mi of shoreline and noncritical on 194 mi. There were 1874 mi of noneroding shorelines. Critical erosion areas were primarily on southeasterly exposed beaches. For example, the west shore of Galveston Bay from April Fool Point to the mouth of Clear Lake at Kemah, a distance of about 10 mi has receded an average of 2 to 4 ft annually.[50] Erosion is generally more severe as fetch (length of open water) increases.

During winter months, approximately 15 to 20 northers will pass rapidly through the coastal area. Rain and winds up to 50 mi (84.5 km)/hr accompany these 24 to 36 hr storms. The north winds generate powerful wave action on the north facing shorelines. Northers tend to "stack up" water on bays initially, causing severe erosion. However, the strong winds cause strong ebb tides that lower bay water levels so that shoreline erosion is shortlived when polar fronts pass through Texas.

Man-induced changes are generally related to the growing populations and industrialization within the coastal zone. For example, in the Chenier Plain there has been a total loss from 1952 to 1974 of 82,080 ha (20.2%) of natural marsh, which includes salt, brackish, intermediate, and fresh marsh. The Chenier Plain of Louisiana is predominantly a rural area. An exception is the Texas portion of the Sabine Basin. The population is generally less than one person per 5 ha as compared to an average of one individual per 3 ha in the rest of Louisiana. Despite the rural character of the area, urban development, along with canals and Gulf waters, replaced 1.5% of the wetlands from 1952 to 1974. About 9.9% of the natural marsh has been impounded. Another 6.4% became inland open water. Spoil areas replace 1.3% of the marsh area, while agriculture accounted for 1.1%.[2]

In contrast to the rural nature of the Chenier Plain, the upper coast of Texas is highly populated and industrialized while the lower Laguna Madre is virtually uninhabited. In 1969, total personal income from the 72 counties of the coastal zone accounted for 40% of the personal income in Texas. Over 170 million tons of cargo passed through Texas ports in 1967.[51] Channels and ports must be kept open by dredge and fill operations to maintain barge and ship traffic.

Dredging and filling activities associated with channels and canals and the associated effects of these operations appear to be major causes of habitat loss or modification. In the Gulf of Mexico, there are 4400 navigation channels in existence, under construction, or planned by the U.S. Army Corps of Engineers. In addition, 138,000 acres of estuaries have been filled.[13] There is probably a 1:2 or higher ratio of canal width to the area required for spoil disposal and levee construction.[52] Thus, the direct acreage loss due to dredging and associated filling is quite high.

Wetlands may be changed directly into open water canals or harbor areas by dredging operations. Spoil dumping after dredging may result in many direct or indirect consequences to marshes. Spoil may be used to fill marshes thus changing them to upland habitat. The spoil may change the hydrologic regime by altering water flow and salinity patterns or segmenting bays by shoaling. Wetlands may be isolated by spoil banks or dikes. Some areas may be maintained as impoundments for waterfowl habitat while others may be drained for agricultural or urban development and mosquito control.[52]

Habitat loss and modifications may be indirectly related to man's alteration of the environmental forces of the coastal zone. For example, channelization may increase saltwater intrusion, increase flushing time, alter tidal exchange, mixing, and circulation, and increase turbidity, which may cause a loss of submerged vegetation. However, channels may be beneficial to estuarine areas by connecting isolated waters and marshes to bays and making them available as fish nursery areas while providing routes of escape or refuge for fish during cold periods. Channels also improve water exchange and circulation while spoil disposal may result in release of nutrients trapped in bottom sediments.[53]

Much of the annual loss of 16.5 mi² (4346 ha) in Louisiana over the past 30 years is attributed to alteration of natural hydrologic and sedimentologic processes. Of this annual loss, 74% occurred in brackish and saline marshes. In Barataria Bay, La., 10% of the wetlands were lost as a direct result of canal construction activities. In Texas, 78,000 acres of fill, primarily spoil from navigation channels, has been placed in estuarine areas. In Louisiana, 26,615 acres of fill has been placed in estuarine areas.[13]

In Texas, about 700 mi of federal navigation channels have altered 13,000 acres of shallow bay bottoms and destroyed 7000 acres of brackish marsh by deepening. Spoil from these channels has covered 23,000 acres of brackish marsh. The amount of destruction of estuarine habitat by private channels is unknown.[53]

Major environmental impacts of industrial and urban areas are effluent discharges, direct habitat loss, and increased runoff of sediments and nutrients into wetlands and streams. Industrial and agricultural toxic wastes occur in significant amount in Texas and Louisiana water basins. The most heavily populated and industrialized basins of the Chenier Plain, Sabine and Calcasieu, have been affected by serious contamination problems from heavy metals and organic pesticides.[2]

In the Chenier Plain, the amount of wetlands freely flooded by tidal waters has been drastically reduced by human alteration of tidal flow. About 162,000 ha are impounded. In addition, many marshes are semi-impounded by canal spoil banks and other levees.[2] Louisiana coastal marshes are often modified by weirs, which are dams constructed 6 to 12 in. below the marsh surface across tidal creeks. The weirs are intended to improve waterfowl and fur bearer habitat, control mosquitoes, prevent erosion, and improve access for hunters and trappers. However, weirs may be detrimental to many fish and crustaceans with salt water affinities.[54] Recent increased rates of land loss in Louisiana's coastal zone are due to control of river flooding, diversion of river sediments, and alteration of wetlands and canals. Sediment from the Mississippi River has historically balanced the effect of subsidence and erosion. Since the early 1950s, levees on the Mississippi River have directed the Mississippi River flow into the modern Birdfoot Delta, thus preventing its normal change to shorter routes, which had resulted in seven historical lobes. The abandoned deltas are in various stages of decay due to subsidence and erosion. Shoreline retreat and alteration are occurring most rapidly in young deltas and slowest in older deltas. Canals built for oil recovery and navigation have accelerated land loss by dredging, filling, changes in hydrology, and increased erosion in channels.[52]

Subsidence is not always associated with deltaic floodplains, Gas, water, and oil removal have caused or accelerated subsidence in many areas. The area of Burnett, Scott, and Crystal Bays near Baytown, Tex. has been an area of severe subsidence due to water, oil, and gas removal. Former land areas are now open water areas or wetland areas.[55]

## IV. SPECIFIC PROJECTS REVIEW

### A. Galveston Island Habitat Rehabilitation Project

The first known habitat rehabilitation project in the western Gulf of Mexico was an attempt at developing the methodology for establishment of aquatic and salt marsh vegetation on barren spoil areas on Galveston Island, Tex.[56] *S. alterniflora* was transplanted in the intertidal zone of a recently established spoil area on Galveston Island to reduce the area's vulnerability to wave erosion and to create an edge effect. Synthetic fibrous material also was placed on piers, bulkheads, and barren shorelines to attract estuarine species. Results indicated that the artificial habitat was more attractive than areas devoid of vegetation. Regretfully, further reports on their transplant and artificial habitat studies were not made.

### B. Shoreline Stabilization with Vegetation in Galveston Bay, Texas

Studies on shoreline stabilization with vegetation in Galveston Bay were initiated in 1973 by the Texas Agricultural Experiment Station through funding by the U.S. Army Corps of Engineers Coastal Engineering Research Center. Vegetation was tried as a cost effective means to control bay shoreline erosion, which had been previously documented by the U.S. Corps Engineers.[50]

The study of shoreline stabilization with native plants evolved into a series of experiments conducted along the shoreline of Anahuac National Wildlife Refuge at the northeast corner of East Bay. Three U. S. Corps of Engineers Coastal Engineering Research Center technical reports were published concerning the work.[57-59] The Ph. D. dissertation by Webb[60] also summarizes most of the work. A summary of the work is presented below.

Bay water salinities were monitored at 2-week intervals at each planting site. Bay water salinities varied considerably. During periods of low precipitation and high evaporation, bay water salinity increased but never exceeded 18.5 $^\circ/_{oo}$. Bay water salinities may have influenced survival of some species, but survival of the predominant intertidal species, *S. alterniflora,* was not affected by these salinity ranges.

Soil texture and chemistry studies were done at the sites. Soil was classified as a loam and clay-loam. Texture varied with elevation in relation to tidal depth and soil depth. Soil salinity and concentrations of cations varied with bay water salinity, precipitation, evaporation, and location of soil in relation to tidal zone, and soil depth.

The initial study involved the transplanting of 12 species of plants with sharpshooter shovels. Except for *Arundo donax* (giant reed) obtained from College Station and *Avicennia germinans* from Galveston Island, all plants were hand dug on Anahuac National Wildlife Refuge. The other ten species were *D. spicata, Phragmites australis* (common reed), *Scirpus americanus* (American bulrush), *S. olneyi, S. maritimus, Spartina alterniflora, S. cynosuroides* (big cordgrass), *S. patens, S. spartinae* (Gulf cordgrass), and *Tamarix gallica* (salt cedar).

Plots were planted on five separate dates and were replicated three times on the south facing shoreline of the refuge. A fourth replication was placed in a shallow and gently sloping ditch that was connected to the bay. The best survival and lowest percent washout of plants occurred at the July transplant date. The plots planted in March, a month of normally gusty south winds, had the lowest survival and the highest percent washout. The significantly higher survival and low percent washout of plants in the ditch vs. that on the shoreline demonstrated that wave action was a severe limiting factor to plants. *S. alterniflora* had the best survival and ranked second in new tillers produced of the species tried. Thus, its value as a shoreline stabilizer was established.

During the first year of the study one area of the shoreline was sloped with a bulldozer and planted. This transplant area was a failure. Most plants were washed out of the middle tidal zone (MHW to MLW). Thus, it appeared that gently sloping shorelines do not attenuate wave forces sufficiently to allow plant survival.

Transplant failure due to wave energy appeared to necessitate establishment of some type of wave-stilling device prior to transplantation. Utilizing readily available hay bales, a hay bale dike was placed at shallow depths along a portion of the shoreline in November 1974. The area was hand planted in November with *S. alterniflora, P. australis, S. cynosuroides, A. germinans,* and *Juncus roemerianus* (needlerush). The 21% survival of transplants behind the hay-bale dike vs. less than 1% in an adjacent unprotected area indicated that wave protection devices helped increase transplant survival. Subsequent observations indicated that this area was relatively high in elevation, with inundation primarily by fresh water due to closeness to Oyster Bayou, *P. australis* demonstrated the best survival rate (4.3%), in both protected and unprotected plots followed by *S. alterniflora* (2.3%), and *S. cynosuroides* (1.8%). The dike was destroyed by waves in March 1975 and all plants subsequently died.

Initial success with the first hay bale wave-stilling device led to construction of a 146-m dike with a larger transplant area. Within 10 days of construction in March 1975 this device was 75% destroyed by wave action. Repairs with 14-gage welded mesh wire and new hay bales lasted only another 5 to 6 weeks. Initially this area appeared to be a failure because of low survival of plants. Monitoring of this area has shown that *S. alterniflora,* despite low survival, has almost completely vegetated this area. Initial transplant establishment appeared to be sufficient to allow slow but steady colonization of the shoreline.

Tire wave stilling devices were placed in front of three areas, which included the sloped area previously transplanted in 1975. All three transplant areas were generally unsuccessful because the tires sank into the sediment allowing wave action over the top of the tires.

In 1976 a second tier of tires was placed over the first tier in the sloped area, which had been unsuccessfully planted on two previous occasions. The second tier of tires was successful in breaking the force of the waves. As a result of the wave protection, excellent survival and tiller production of *S. alterniflora* occurred. Density and height of *S. alterniflora* increased with increasing water depth and hours of inundation. Good survival of *S. spartinae, S. patens,* and *D. spicata* occurred just above MHW. Regardless of species, infrequent inundation and concomitant dry soil conditions limited survival at high elevations. *J. roemerianus* transplants failed to survive in significant numbers, regardless of hours of inundation.

Due to a 0.15 m buildup of sediment behind the wave barrier, removal of one half of the wave barrier in the sloped area in 1978 resulted in 6 m of landward erosion over a 1.5 year period. After the initial erosion ceased, *S. alterniflora* began recolonization outward, thus, demonstrating that temporary wave-stilling devices can be successfully used.

The studies along the shoreline of Anahuac National Wildlife Refuge in East Bay, Tex. indicated that:

1. Transplants along shorelines exposed to a long fetch will not likely survive unless wave protection is provided.
2. Temporary wave-stilling devices, that can be removed after plant establishment, will work when properly installed.
3. Proper installation of wave-stilling devices should reflect an awareness of the tendency for the barrier to sink into the sediment and the tremendous force of wind pushed waves.
4. Wave barriers made of two tiers of tires held in place by cable and steel posts were successful and relatively inexpensive.
5. The intertidal zone from MHW to low water appears to be the critical erosion zone, normally; thus, *S. alterniflora* is the critical species to establish.
6. A number of species may be used above normal MHW to prevent erosion by wind pushed tides and storm tides.

Shoreline characteristics along MHW, such as elevation, slope, and soil texture, will determine the species to be selected for shoreline stabilization. The normal zone of occurrence and successional characteristics of *D. spicata, S. patens,* and *S. spartinae,* along with results of transplant studies, indicate these three species to be reliable transplant species under varied conditions.

### C. Bolivar Peninsula Habitat Development on Dredged Material

During 1976, an investigation to test the feasibility of developing marsh and upland habitats on dredged material was initiated on Bolivar Peninsula, Tex. The study was funded by the dredged material research program managed by the U. S. Army Corps Engineers Waterways Experiment Station. The experimental planting and sampling were performed by the Range Science, Wildlife and Fisheries, and Soil Chemistry Departments of the Texas Agricultural Experiment Station.[27]

Two grass species, *S. alterniflora* and *S. patens,* were hand planted with sharpshooter shovels in the intertidal area. Except for a control area, plantings were protected from wave energies by a sandbag dike. Plant survival and growth in response to fertilizer treatments were monitored through 1978. Follow-up studies were conducted through 1980.

Transplants (sprigs) of *S. alterniflora* that were protected from wave energies survived successfully throughout the marshland. However, survival, density, and height of plants generally were greatest at lower elevations and decreased with increasing elevation to approximate MHW. High interstital water salinity at MHW apparently affected plant survival and growth. Above MHW, survival, density, and height increased with increasing elevation to the maximum planting elevation of 2.5 ft.

Transplants of *S. patens* failed to survive below approximate MHW. *S. patens* survival and establishment increased with elevation above MHW. This species was better adapted to areas above MHW than *S. alterniflora* was.

Establishment of seeds of both species planted in March 1977 occurred only at elevations above the approximate MHW level. The March seeding appeared to be poorly timed for the experimental planting despite favorable temperatures. Lack of seedling establishment at lower elevations was attributed to high wave energy and suspended silt, which affected the amount of light available to plants. Sinking of sandbags into the sediment allowed wave action over the top of the bags to be a factor in seed germination and survival. Natural seed germination and establishment during December and January at all levels of the marshland indicated seed establishment is possible without high tides and wave energies.

Soils analyses before planting indicated that total kjeldahl nitrogen, exchangeable ammonium, extractable phosphorus, cation exchange capacity, organic matter, and percentage silt and clay were generally low in this sandy substrate. All elements decreased with increasing elevation. Despite the low measurements, fertilizer treatments did not significantly enhance plant growth and may even have been detrimental to initial plant survival.

After planting and fertilization, the same elevational trends remained but concentrations of organic matter, nitrogen, and phosphorus generally increased over time. The major factor contributing to changes in soil characteristics was the rate of sedimentation. The sandbag dike lowered wave energies, which caused deposition of coarse grained material in a berm between the sandbag dike and plots and 15 cm of fine-grained sediment in the plots. The greater biomass of *S. alterniflora* at lower elevations may have indicated a response to the greater availability of nutrients at lower elevations.

Termination of the project eliminated observation of aquatic biota at a time when the marsh was developing. Preliminary observations indicated that with marsh development, benthic organisms, insects, and fish populations increased. Bird diversity and activity in the planted marsh also increased with marsh development.

The work on Bolivar Peninsula indicated that wave protection to plants allowed marsh development. However, it is recommended that a more cost effective wave break be used rather than a sandbag dike. Fences may be necessary to protect sites from large grazing animals. *S. alterniflora* should be sprigged at elevations below mean high tide, whereas *S. patens* should be sprigged at elevations above mean high tide. Seeding may be successful with low tides and low wave energy. Fertilization does not appear necessary. Species diversity and numbers of benthic organisms, aquatic organisms, birds, and mammals appeared to increase with marsh development.

### D. Stabilization of Dredged Material in Corpus Christi Area

*S. alterniflora* was transplanted on the shoreline of Rincon Industrial Park and Harbor Island in Corpus Christi Bay.[61] At least 80 transplants from each of several sources were utilized on each of three transplant dates. *S. alterniflora* was successfully transplanted, but seasonal factors mitigated the level of success. Fluctuating water regimes, temperature, interstitial water salinities, phenological phase of plant development and elevation above and below MLW were important factors in transplant success.

The June planting at Rincon was successful in the lower tidal zones. The August planting was unsuccessful. High temperatures, low precipitation, poor plant stage, and high interstitial water salinities apparently were damaging to plants. At these two dates, the source of transplants appeared to make no difference. During the November transplanting at Rincon, the success of each plant source was variable. Time of collection and time of storage prior to planting may have been a factor. Plantings in the tidal zone were the most successful.

At Harbor Island, local transplant material transplanted more successfully during the November planting than did other sources. Survival occurred only in the tidal zone, with better success with increasing water depth.

## E. Revegetation of Pipeline in Texas and Louisiana Wetlands

Revegetation of 11.9 mi of wetlands along the Galveston-High Island Pipeline in southeastern Texas and northwestern Louisiana was performed by Transcontinental Gas Pipeline Corporation and monitored by Chabreck.[62] The pipeline originated offshore, crossed into Texas west of Sabine, and passed into Cameron Parish, La. The objectives of the revegetation program were to restore vegetation within disturbed areas, to compare natural revegetation rates in single- and double-ditched areas, and to develop alternate revegetation methods to augment natural revegetation. Marsh types varied with elevation and salinity regimes. The marshes included low brackish marshes, high brackish marshes, low salt marshes, and high salt marshes.

The ditch was 8 ft wide and 5 ft deep. Soil from the ditch was placed beside the ditch between the time of excavation and backfilling. Two storage areas were used with the double ditch method. The top 6 to 12 in. of soil with vegetation was placed in one area while the subsurface soil was placed in a separate area. With the single-ditch method, the soil removed from the ditch was not separated by depth. Test plantings were made in the ditched portion of the construction area in February 1978. Plantings of *D. spicata* and *S. patens* were made in the brackish marsh type. *D. spicata* and *S. alterniflora* were planted in salt marsh areas. Six plots of each species were planted in each marsh type. Each plot of the designated species contained four sprigs consisting of a cluster of five stems and associated roots.

Vegetation reestablished naturally in all plots by August 1979. In low salt marsh plots, plant density and cover approached that of undisturbed areas. However, in high marsh plots in Texas, regrowth was less than 50% of that of disturbed areas. In Louisiana, natural vegetation had recovered satisfactorily. However, establishment of perennial plants in spoil storage areas was hindered by slightly greater water depths than undisturbed areas. The greater water depths probably resulted from soil compaction and subsidence from the weight of the stored material or from excessive raking by the dragline crew while gathering backfill. This condition may actually have been beneficial to wildlife because annual plants, which provide good food, colonized the lower areas.

Transplanted stems had good survival and growth in most plots with the number of stems increasing within 6 months to over five times the number planted. However, natural regrowth occurred at a greater rate. Chabreck[62] concluded that planting had no significant effect on the recovery rate of the vegetation.

The double-ditch method had slightly greater revegetation rates than the single-ditch. Plots in three of the four marsh types produced greater vegetative cover where the double-ditch method was used.

## F. Texas City Dike Marsh Establishment

In early 1979 a marsh establishment project, which was a prototype for a future 600 acre area, was initiated along the Texas City Dike in Galveston Bay.[63] The purpose of the project was to demonstrate that a salt marsh could be established on maintenance dredged material. The project was planned and funded by the U. S. Army Corps of Engineers Galveston District and directed by R. J. Bass.

Wave protection to plants was provided by a hydraulically pumped levee of clay, which had been taken from the edge of the Texas City ship channel. A bulldozer was used to slope the levee. Despite construction of the levee to a 4 ft height above MHW and an approximate 12 ft width, erosion was severe due to the 20-mi fetch to the north. A near breach of the levee due to wind generated waves from tropical storm Claudette several months after construction dictated immediate use of rip rap. However, the levee continued to erode even after the placement of rip rap. By May 1981, the elevation was close to MHW and less than 6 ft wide.

A slope with a median elevation near that of natural marshes of the area was established. However, considerable problems were encountered in establishing the slope.
The 30 in. dredge was so large that the dredged material, which was fine-grained,
flowed across the area and out the spillway in the levee. Several modifications to normal pumping procedures were established to fill the area. The dredge pump was slowed
as much as possible. The discharge was split so that only half flowed into the levee
area. A temporary elevated spillway was built. After the area was filled, the material
was too fluid to allow it to be sloped. To achieve the desired slope the area was pumped
6 in. higher than the desired depth and allowed to compact for 6 weeks. Following
compaction, an amphibious dragline was used to cast material from outside the levee
onto the area adjacent to the Texas City Dike. Material from the channel was then
pulled over the new material. The temporary spillway was removed to allow settling
of the spillway end.

Planting of the area was done in early May 1979. Plant spacing varied from 0.5 m
between plants in one third of the area to 1m in one third of the area. The other third
was not planted. Since the substrate was too soft to walk on, a boat with a planting
crew was pulled back and forth across the area. Three persons in the boat inserted
plants into the mud. Five rows were planted with each pass of the boat. Three days
were required to complete the planting. This procedure worked well.

The night following the completion of planting, 31 mi/hr winds caused severe wave
turbulence across the bay. Due to the wave protection provided by the levee and the
Texas City Dike, only an approximate 10% of the plants were uprooted. Approximately 90% transplant survival occurred.

By October of the first year, the 0.5 m and 1 m plant spacing areas looked similar.
Plants were as tall and vigorous as natural marshes of the area. However, plant density
did not appear to be equal to the natural marshes. The following February, seedlings
of *S. alterniflora* appeared throughout the area. With maturation of seedlings and
further growth of original transplants, the area resembled a natural marsh by the end
of the second growing season.

Bass[63] concluded from his work that: (1) marsh establishment in a similar manner
is feasible on the 600 acre area that will be established in the future, (2) a permanent
rip rapped dike will be necessary, (3) a 1 m spacing of plants was sufficient and (4)
the entire area need not be planted since seedlings would invade open spaces left
throughout the area.

## G. Calcasieu Lake and Houma, Louisiana Transplantings

Attempts at establishment of stands of *S. alterniflora* for future planting stock were
made May and June 1981 in Calcasieu Lake, La. and in marshes around Houma, La.[64]
Both areas were predominantly fresh water wetlands until saltwater intrusion occurred.
Establishment of stands of *S. alterniflora* may now be possible with the increasing
salinity. In Calcasieu Lake this species may provide a means of combating shoreline
erosion. Plants were spaced on 10 ft centers in three parallel rows. Plants were placed
in wave protected areas 8 to 10 ft back of the shoreline. Plants that were indigenous
to the area and commercial nursery stock from Florida were tried. A slow release
fertilizer tablet of 27-7-4 strength was placed in the hole for each plant. The cost was
$0.08 for each tablet with a calculated rate of 500 lb of nitrogen per acre. Time to dig
and transplant material was recorded but results of this work are not yet available.

## H. Marsh Establishment on Dredged Material, Steadman Island, Texas

A contractor (Charles Wolf of Houston) for the U. S. Army Corps of Engineers
planted *S. alterniflora* on a dredged material site at Steadman Island near Aransas
Pass, Tex. in May to June, 1981.[65] The site was planted as mitigation after the dredged

material flowed into wetlands of Corpus Christi Bay. During the restoration, a levee was constructed to retain the dredged material. After overflow material was placed back into the leveed area, *S. alterniflora* was hand planted with sharpshooter shovels at 1 m intervals into 13 acres of the area. Elevation zone of planting along a gradual slope was supposed to be between 2.5 ft and 1.5 ft above mean low tide. Because continuous high tides occurred during the planting contract period, planting elevations were shifted up the slope. Plants were dug from marshes in the local area, placed in pickle buckets for transporting, and planted as single culms soon after digging. A wheel designed to punch a small hole at 1 m spacing was used to establish spacing of plants in each row. Fetch at the site did not appear to be more than a couple of miles. The depth of the bay was shallow with a very gradual slope.

The contract bid on planting the area was $1700/acre. Observations were made by the author on June 26, 1981. Excellent survival was achieved at lower elevations. Most plants above the 2.5 ft contour were dead.

### I. Marsh Mitigation for Habitat Loss at Clear Lake, Texas

A salt marsh was established in 1977 in the town of Nassau Bay, Tex.[66] The area is located in front of Bay Front Apartments on the shoreline of Clear Lake. Marine, Inc., under the direction of Josh Tillinghast Architects, established the marsh behind an existing flow-through bulkhead. The area behind the bulkhead had been filled, but it was lowered to an elevation equivalent to that of local salt marshes. A stone retaining wall was constructed to prevent erosion of adjacent upland into the site. *S. alterniflora* was dug from nearby marshes and placed for several hours into a root stimulant as more plants were being dug. The root stimulant was Alginure®, which is made from kelp and only available in Germany. The concentration was 2 oz for each 5 gal of water. Plants were inserted into the soft mud at 36 in. spacings between plants. A broad spectrum fertilizer was broadcast over the 25 by 585 ft area.

This salt marsh successfully established. Cost of establishment is unknown since marsh planting was not separated from other items in the overall budget.

### J. Lake Pontchartrain Shoreline Erosion Demonstration Project

In November 1979, 1900 ft of shoreline on Lake Pontchartrain in Louisiana were planted with *S. alterniflora, P. australis, S. patens,* and *Rosa bracteata* (MaCartney rose).[67] There were 14 plots in the intertidal zone and 14 plots in the beach zone. Each plot was 120 ft long. The primary objective of the project was to stabilize the sand beach to keep it from washing back into a wetland dominated by *S. patens*.

Peat pot seedlings, which were grown in a local nursery, and sprigs of *S. alterniflora* were planted. However, after an adequate supply of plants was located, seedlings were no longer required.

Several problems were encountered during transplanting. Cattle had to be fenced out of the transplant area. Plants were repeatedly washed out of the sandy soil by wave action, which was generated over a 20 mi fetch. Since the water depth was shallow, less than 1 to 2 ft in depth, waves sometimes reached 3 to 4 ft in height.

Wave protection devices of sand cement bags, tires, and brush fences were tried, but they were not successful. Plants were able to establish as long as root systems penetrated into the organic soil that was present. A 10 ft wide area of vegetation eventually established along the shoreline in the peat soil. This area was originally 6 rows of plants at 18, 24, and 36 in. spacing.

Although *S. alterniflora* established on the shoreline, the stabilizing effect on the sand was questionable. Since plants did not grow in the sand, the sand still continues to be pushed across the wetland.

## K. Natural Revegetation of Dredged Material and Scraped Areas

Many areas of bare dredged material or disturbed sites occur throughout Texas and Louisiana. Various studies have shown that by construction to the proper elevation, some areas will colonize quickly with desired vegetation if wave action is not severe or elevation is above normal wave action.

In Texas, four salt marshes were altered in violation of Section 10 of the Rivers and Harbors Act and Section 404 of the Clean Waters Act, which are enforced by the Corps of Engineers. After restoration of each site to the proper elevation these sites were able to colonize with *S. alterniflora* and other salt marsh vegetation over a period of years. Most species that were able to colonize were upper elevational salt marsh species such as *Salicornia bigelovii, S. virginica,* and *Batis maritima.*[68]

Along the Southwest Pass in the Mississippi River Delta, La. elevations of the vegetative cover were surveyed in unconfined dredged material disposal areas. *Eleocharis parvula* (dwarf spikerush) was chosen as the major indicator species for tidally flooded marsh areas. Other species used as secondary indicators in the marsh were *Scirpus americanus, S. maritimus, Zizaniopsis miliaceae* (giant cutgrass), *Colocasia antiquorum* (elephants ear), and *Sagittaria platyphylla* (delta duckpotato). The upper limit of marsh created from release of unconfined disposal material is 1.92 ft mean sea level. This data is important in ongoing studies and in actual creation of marshes with dredged material. Natural colonization of unconfined dredged material will convert shallow bay bottoms along Southwest Pass into fresh to intermediate marsh types and provide habitat to wildlife.[69]

## V. RECOMMENDED TECHNIQUES

### A. Species Selection

In saline intertidal areas the plant species selection seems to be limited to one species, *S. alterniflora.* The normal growing zone for this species is from MLW to approximate MHW. No other species in the western Gulf of Mexico marshes can tolerate the high salinities and frequent inundations of the lower intertidal zone of salt marshes and estuaries. Fortunately, this species transplants well and spreads rapidly under normal growing conditions.

At elevations above normal high water a number of species can be utilized. The species to select will depend on many environmental factors, such as salinity regime, elevation, and soil moisture regime. The species selected should match the environmental characteristics of the area. The typical zonation of salt marsh plants suggests that *D. spicata* and *Spartina patens* should be used at elevations frequently inundated with salt water. Lower salt marsh elevations with low salinity levels may be transplanted with *P. australis, Scirpus americanus, S. olneyi,* and *Spartina patens.* Higher elevations may be planted with *S. spartinae.*

### B. Digging of Plants and Transplant Material

Plants are generally dug with a shovel because this is the simplest method. This does not negate the possibility of using backhoes or plows. Plants may be pulled from the ground sometimes but one should be certain that adequate roots are obtained with the plants.

The characteristics of each plant species will dictate transplant material to be used. *S. alterniflora* is generally planted as a single main culm (stem) with or without smaller tillers emerging from the stem base. Single culms, which normally survive well, allow large areas to be transplanted with a minimum amount of transplant material. Large clumps of *S. alterniflora* have been transplanted with great success and there are some indications that large clumps can withstand wave action better than single culms.

Species other than *S. alterniflora* must be handled differently to obtain good transplant material. *S. patens* and *S. spartinae* separate readily into bundles of 3 to 7 stems with their associated roots. *D. spicata* transplant material seems to vary somewhat from area to area. Generally, a small plug of sod or a handful of roots with its associated stems will suffice. *Scirpus americanus* and *S. olneyi* transfer best as several stems with associated roots. *P. australis* is planted as one stem, with the top clipped, and its associated roots.

Soil can be knocked or washed free from plant roots to facilitate handling and movement. However, care should be taken to keep roots moist between the time of digging and transplanting. Placement of wet burlap bags over the lower portion of plants, placement of plants into buckets with several inches of water, or heeling-in (covering roots with soil) are acceptable techniques for the temporary storage of plants.

Use of root stimulants is of questionable value. Alginure® was used successfully in one transplant area but no comparison was made to plants that were not treated. Tanner[70] found that the root stimulant he used was very detrimental to plants. Plants should, ideally, be planted immediately, but holding plants overnight has been successfully done.

The location of digging does not appear to be critical. Plants from the local area may be better adapted to local environmental conditions than plants from other areas or from nurseries. Nursery stock may be easier to acquire and use than plants from native stands. However, local transplant material is usually quite economical to obtain. The time and cost of digging can be minimized with prior planning and experience. One shovelful from a good stand of *Spartina alterniflora* should produce multiple stems. If only one stem is being pulled or dug at a time, a new site should be located to conserve time and energy.

Sandy areas, rather than silty areas, are generally much easier to dig and walk in. New growth areas, such as the edge of established stands, generally produce multiple stems with each dig of the shovel.

## C. Transplanting

Small plants of *S. alterniflora* are easier to handle and tend to withstand wave energies better than tall plants, which have more area exposed to waves. One precaution should be adhered to as far as size. If short plants of *S. alterniflora* are placed at lower tidal elevations, chances of survival are diminished since silt in the water may block light while tides may not drop low enough to allow exposure of leaves to sufficient light. Thus, plants at lower elevations may not survive.

*S. alterniflora* can be simply pushed into soft mud substrates. Sharpshooters, pine seedling divots, or other shovel types are the simplest instruments for transplanting in firm substrate. Simply insert the shovel into the ground, pry forward and back to create an opening, insert the plant to the base of the stem, remove shovel, and pack firm with a foot.

Working in pairs with one person digging and the other planting seems to be more efficient than one person digging and planting. The main problem in planting is the caving in of the hole in wet or inundated areas. With fluid soils it probably will be necessary to insert the plant as soon as the hole is opened and pack the soil firm before the plant is released. By postponement of planting until low tide or until suitable soil conditions exist, considerable time and effort per transplant can be saved.

Experiments in the western Gulf of Mexico indicate that fertilization is not necessary during transplantation. However, if fertilizers are applied, then the following recommendations are suggested. Slow release fertilizer pellets or tablets, apparently, can be placed directly into each hole as transplanting occurs. Soluble fertilizers such as ammonium sulfate or ammonium nitrate should not be placed directly on plant roots. Broadcast fertilization may be necessary if the fertilizer cannot be incorporated into the soil.

Depth of the hole should vary with each plant since bending of roots (J-rooting) should not occur. Average depth of each hole will be about 4 in.

Boats or sleds can be used on soft mud to facilitate transplantation. The persons in or on the boats or sleds should position themselves to insert the plants as the boat or sled is pulled across the mud.

Large areas can be economically planted by a mechanical planter, provided the substrate will support a tractor or tracked vehicle, which is needed to pull the mechanical planter. Mechanical planters can be purchased from a number of sources for about $500. Although self-propelled machines are available, one designed to attach to a three point hitch of a tractor is most feasible. Spacing of plants can be adjusted. One major problem of mechanical planters is that tall plants tend to catch on the sprockets. Therefore, trimming of plants may be required. Effects of trimming of plants has not been well tested but it does appear desirable to leave as much of the leaves on the plant as possible. Mechanical planters work best with no tidal inundation at the time of planting. Obtaining, transporting, and planting of plants under the ideal conditions takes some prior planning and knowledge of the times of daily tides. A little luck may be necessary to avoid delays due to storm tides.

### D. Seeding

*S. alterniflora* seeds can be collected in late autumn by clipping the seed heads of plants. Seeds should not be collected until seeds become mature (firm seeds past the dough stage). In Texas, the time of collection is generally the latter part of October. However, different stages of development of seeds occur by locality and elevation.

After collection it is necessary to store the seeds at 1 to 2°C (just above freezing) in seawater or artificial seawater. Seeds must go through an after-ripening period of a month or two before germination will occur, but prolonged storage of 6 months will result in loss of viability of seeds. Therefore, collection each autumn is necessary if seeding in the following spring is desired.

Seeds can be utilized to establish potted seedlings and nursery areas, or they can be planted directly into the transplant area. A shallow covering of soil is desirable for best germination and seedling establishment.

*S. patens* seeds can be collected during late summer and during the autumn. These seeds should be stored dry until time of planting.

### E. Elevation Requirements

Generally, *S. alterniflora* transplants and seeds should be placed in the intertidal zone from MHW to MLW. Best survival and growth of transplants will generally be at the lower tidal range. *S. alterniflora* seeds may not survive well at lower tidal ranges during periods of frequent complete inundation.

*S. patens* and other transplant material will generally be placed above *S. alterniflora*. However, the elevation of growth of *S. alterniflora* and *S. patens* may not be easily recognized on some shorelines. Alternating rows of each species at the questionable elevation allows the best adapted species to colonize.

The elevation to plant can be derived from local *S. alterniflora* marshes on calm days. Observe the water level in the local marshes, proceed immediately to the proposed transplant location, and mark the tide level. The upper and lower limits of transplanting can be estimated by appropriate measurements in the natural marshes. The best measurements can be made when tide level is at the upper edge of local *S. alterniflora* stands. With a meter stick, the depth to the lower level of growth can be measured and transposed in a similar manner to the proposed marsh area.

### F. Time of Transplanting

Transplantation can be done during any month with reasonable success if wave conditions at the site are not too severe. Best survival of transplants is generally during

59. **Webb, J. W. and Dodd, J. D.,** Shoreline plant establishment and use of a wave-stilling device, Misc. Rep. No. 78-1, U. S. Army Corps of Engineers, Coastal Eng. Res. Cent., Fort Belvoir, Va., 1978.

60. **Webb, J. W.,** Establishment of Vegetation for Shoreline Stabilization in Galveston Bay, Texas, Ph.D. thesis, Texas A&M University, College Station, 1977.

61. **Oppenheimer, C. H. and Carangelo, P. D.,** Biological applications for stabilization of dredged material Corpus Christi area, Six Month Prelim. Rep., University of Texas Marine Science Institute, Austin, 1978.

62. **Chabreck, R. H.,** Revegetation program for pipeline construction areas in certain Texas and Louisiana wetlands, Final Rep. to Transcontinental Gas Line Corporation, Houston, Tex., 1979.

63. **Bass, R. J.,** personal communication, 1981.

64. **Cutshall, J.,** personal communication, 1981.

65. **Anthamatten, F. L.,** personal communication, 1981.

66. **Tillinghast, J.,** personal communication, 1981.

67. **Dement, L.,** personal communication, 1981.

68. **Pitri, R. L. and Anthamatten, F. L.,** Successful restoration of filled wetlands at four locations along the Texas coast, in *Wetlands,* 1, in press.

69. **Montz, G. N.,** A vegetational study conducted along Southwest Pass in the Mississippi River Delta, Louisiana, *U. S. Army Corps Engineers, New Orleans District,* 1977.

70. **Tanner, G. W.,** Growth of *Spartina alterniflora* within native and transplant-established stands on the upper Texas Gulf Coast, Ph.D. thesis, Texas A & M University, College Station, 1979.

Chapter 5

## PACIFIC COASTAL MARSHES

**Paul L. Knutson and W. W. Woodhouse, Jr.**

TABLE OF CONTENTS

# I. INTRODUCTION

## A. Natural Plant Communities

Pacific coastal marshes are less uniform than those found on the Atlantic coast of the U.S. The vast majority of marshes on the Pacific are found in scattered bays and lagoons and in the mouths of some tributaries. The coastal marshes of central and southern California are very distinct from those in the Pacific northwest. Along the southern Pacific, seasonally high levels of salinity in coastal estuaries greatly limit the diversity of intertidal vegetation. In portions of the larger bays such as San Francisco, salinity concentrations reach or exceed sea strength during the dry summer months when evaporation exceeds rainfall runoff and tidal exchange. Smaller tributaries periodically run dry allowing sea water to intrude upstream. Under these conditions Pacific cordgrass (*Spartina foliosa*), the west coast equivalent of the Atlantic smooth cordgrass (*S. alterniflora*), is the dominant flowering plant of the regularly flooded portion of the intertidal zone (Figure 1). Only in the upper reaches of tributaries where fresh water is a consistent influence is cordgrass displaced by other species such as Alkali bulrush (*Scirpus robustus*). Broad reaches of the irregularly flooded high marsh are dominated by pickleweeds (*Salicornia* spp.), a plant widely distributed, but of little importance on the Atlantic coast. The stresses of salinity in the high marsh are even more severe than in regularly flooded marshes. The ability of pickleweed to tolerate salinity concentrations of more than twice sea strength provide it with a formidable advantage over most other species. Saltgrass (*Distichlis spicata*) is common but seldom a dominant high marsh species.

Fresh water is a much greater influence on the marshes of the Pacific northwest. In some areas such as the Olympic Peninsula of Washington, annual precipitation is over 250 cm/year. Specific composition of these marshes is more diverse than that on other coasts of the U. S. and elevational "zones" are less discrete. There is no single species such as Pacific cordgrass colonizing and dominating the lowermost, regularly flooded zone of this region. The regularly flooded low marshes are characterized by Lyngbye's sedge (*Carex Lyngbyei*), tufted hairgrass (*Deschampsia caespitosa*), spike rushes (*Eleocharis* spp.), pickleweed, three-square bulrush (*Scirpus californicus*), Baltic rush (*Juncus balticus*), and seaside arrowgrass (*Triglochin maritima*) (Figure 2). As in the southern Pacific coast, saltgrass is common in the high marsh but is seldom dominant.

## B. Historic Losses

In general, Pacific marshes are formed in relatively narrow river mouths which drain almost directly onto a steeply sloping continental shelf along a slowly emerging coastline.[1] They are far less abundant than the coastal marshes of the Atlantic. On the Pacific, California has the most extensive salt marshes, about 36,000 ha.[2] Washington state has only about 4500 ha, while Oregon has about 3040 ha.[3]

California has the dubious distinction of being the Nation's leader in the destruction of marshes and other wetlands. Two thirds of California's wetlands have been diked or filled in the past century.[4] In San Francisco Bay, the states greatest concentration of coastal marshes, nearly 60,000 ha have been destroyed. In southern California, only Morro Bay retains a significant portion of its tidelands. At one time, San Diego's Mission Bay supported nearly 400 ha of mudflat and marsh. Today, less than 20 ha of marsh and virtually no mudflat areas remain.[4]

## C. Values

Only recently has there been widespread recognition of the critical value of coastal marshes as feeding and nursery areas for fish and birds, as a source of energy and oxygen for marine organisms, as sinks for nutrient and metal pollutants, and as stabilizers of eroding shorelines. Most species of sport and commercial fishes spend at least a portion of their life cycles in estuaries where they depend on marshes. Marsh vegetation, phytoplankton, and seagrasses are the crops of the sea. They are the vehicles by which the sun's energy is ultimately converted to animal protein.[5] Most research on this subject, however, has focused on Atlantic coast marshes. Only recently has there been comparable work on the productivity of Pacific marshes. Eilers[6] found primary productivity in an Oregon marsh comparable to Atlantic marshes. Production varied from about 5000 to 20,000 kg/ha. Higley and Holton[7] found food chain relationships in two Oregon marshes similar to those reported on the Atlantic.

The major destruction of coastal marshes, particularly in California, has undoubtably reduced the natural abundance of the coastal zone. San Francisco Bay no longer supports a commercial crab or shellfish industry and other fisheries resources have declined substantially.

## D. Marsh Restoration

The Pacific coast is a prime candidate for marsh restoration. A goal has been expressed by the California Legislature to restore 25% of the state's lost wetlands by the year 2000. There have been far fewer restoration studies conducted in the west. In the 1970s, a series of experimental plantings were made in San Francisco Bay.[8-12] Additional plantings were made in the Pacific northwest.[13]

The guidelines presented in this chapter are based upon both experience and current literature. Existing information will require considerable extrapolation and many of the resulting recommendations will be speculative. Woodhouse[14] has summarized planting guidelines for coastal marshes of the U. S. His report is the primary source of information for this chapter though additional information has been taken from Knutson[15] and Knutson and Woodhouse.[16]

## II. PLANT MATERIALS

This chapter emphasizes the plants that, at present, appear useful in coastal marsh creation. The small number of these plants is due to the saline conditions prevailing in most coastal marshes, the rigorous conditions during establishment, the difficulties encountered in propagating some species, the lack of information available on others, and the secondary role a number of them play in the marsh environment. A great deal is known about where and under what conditions many marsh plants grow. Interest in planting them is of recent origin and planting requirements are known for only a few. More species are likely to be found useful in the future.

In this section, six coastal marsh species will be discussed individually; Pacific cordgrass, pickleweed, saltgrass, Lyngbye's sedge, tufted hairgrass, and seaside arrowgrass. In general, the first three species will be used for marsh restoration on the southern

FIGURE 1.    Pacific cordgrass marsh.

FIGURE 2.    Lyngbye's sedge marsh. (Photo courtesy of D. L. Higley and R. L. Holton.)

Pacific coast from Humboldt Bay, Calif. south while the later three species will be used primarily in the states of Oregon and Washington. A few additional marsh plants that are found on the Pacific coast will be discussed briefly at the end of this section.

### A. Pacific Cordgrass (*Spartina foliosa*)
*1. Plant Description*

This grass is similar in appearance to the smooth cordgrass of the Atlantic and Gulf coasts but does not grow quite as tall and is less vigorous and slower to establish. Pacific cordgrass (Figure 3) is the dominant flowering plant in regularly flooded intertidal marshes from Humboldt Bay southward to Mexico.

Two forms of Pacific cordgrass have been identified in San Francisco Bay; a medium, stout form (0.3 to 1.2 m high) which grows in the lower zone and a dwarf form (0.2 to 0.3 m high) which occurs mixed with pickleweed (*Salicornia* spp.) at higher elevations.[17] It is not known whether these forms have a genetic basis or are due to

FIGURE 3.    Pacific cordgrass — San Francisco Bay, Calif.

environmental factors. Short-term field tests suggest that the two forms react differently to elevation. The dwarf form was able to survive transplanting in higher zones than was the stout form.[18]

Reproduction in established stands is vegetative through extensive underground stems (rhizomes). Seed production is erratic and usually limited in old, dense stands, but it may be substantial in newly established stands or along the margins of older stands. Seeds are important for spreading the plant into barren or disturbed areas.[17]

Pacific cordgrass can be established in either sand or fine-grained sediments. However, it is more likely to be nutrient-limited in sandy substrates.[19]

Purer[20] observed cordgrass in saline environments from 22 to 30⁰/₀₀. Floyd[21] found germination rates higher in fresh water than in salinities of 10, 20, and 30⁰/₀₀. Phleger[22] subjected adult Pacific cordgrass plants to salt solutions of from 0 to 40⁰/₀₀. He found that growth was best in fresh water. It appears that plantings can be made in salinities up to about sea strength (35⁰/₀₀).

Submergence by the tides is probably the most important environmental factor affecting the distribution of intertidal plants. Pacific cordgrass is remarkably well-adapted to withstand long periods of inundation. Most plants exchange gases (breath) through small openings in their leaves known as stomata (from the Greek meaning "mouth"). In Pacific cordgrass the stomata are sunken and the "lip-like" guard cells which surround the stomata are accompanied by subsidiary cells equipped with

FIGURE 4.    Seed collection. (Courtesy of San Francisco Bay Marine Research Center.)

branched papilla (finger-like projections). It is suggested that these papilla trap air bubbles and prevent the wetting of the stomatal apparatus during submergence.[23] Like several other members of the genus *Spartina,* Pacific cordgrass has large air spaces within its shoots and roots. These air spaces (parenchyma tissue) allow the plant to store oxygen for respiration during submergence.[20,23] Experiments have also demonstrated that oxygen is transported downward through these tissues to the plants subsurface roots and rhizomes.[24] This adaptation may allow the lower portions of the plant to carry on respiration and exchange of gases via the emergent stems even when the plant is partially submerged. Because of this ability, cordgrass survives lower in the intertidal zone than any other emergent plant in its range. For all practical purposes, planting efforts using Pacific cordgrass can use mean tide level or slightly below as the lower boundary for successful establishment. The upper boundary is about mean higher high water.

### 2. Cultural Techniques

Pacific cordgrass is propagated by seeds, by vegetative materials harvested from natural stands, or by vegetative materials produced in nurseries.

### a. Seeding

Seed production in Pacific cordgrass is very erratic. Early investigators believed that viable seeds were seldom produced and of minor significance in the spread of this species.[20,25] However, recently substantial seed crops have been collected and germinated in San Francisco Bay. The highest seed production occurs near tributaries. The lower salinities at these sites may be a factor although this has not been demonstrated experimentally. Pacific cordgrass seed heads may be attacked by an ergot (*Claviceps purpurea*).[4,17]

Pacific cordgrass seeds mature in the San Francisco Bay area in October; seed heads begin to shatter shortly thereafter. Harvesting must be timed just before shattering when some seeds are easily dislodged by tapping the heads. Mature heads may be clipped by hand either from a boat or by wading (Figure 4). Seeds should be stored in cold saltwater for about 2 weeks to loosen inflorescences. Seeds may then be threshed by placing heads on a No. 30 screen and subjecting them to a strong spray of water from a hose. Viability of seeds has been maintained over winter by storing in cold (4°

C) fresh or salt (12⁰/₀₀) water. Saltwater is more effective in preventing germination during storage. Satisfactory germination has resulted when seeds were placed in freshwater at the end of the storage period. There is some indication that germination of Pacific cordgrass may be more sensitive to salinity than smooth cordgrass. Viability is not maintained by drying or freezing.[4,10,17]

Plants have been produced from seeds by direct seeding in sand, sand-silt, or vermiculite mixtures in peat pots or by germinating seeds in petri dishes and transplanting to peat pots.[17]

There is considerable doubt concerning the dependability of using seeds to produce stands of Pacific cordgrass. Vegetative materials have been consistently more successful.

### b. Field Harvesting

Sprigs are the least expensive plant material and are easy to transport and plant. A sprig of Pacific cordgrass consists of at least one stem with attached root material. Sprigs are obtained from existing marshes. Presumably, field nurseries can be established as a source of plant materials, though to date, this has not been practiced. In sandy substrates individual clumps may be loosened with a shovel, lifted, and separated into individual sprigs. This task is much more laborious and likely to produce poor quality plants in fine-grained sediments. In sandy sediments the highest quality transplants can be obtained from uncrowded stands which do not have a dense root mat. Because the natural spread of Pacific cordgrass is relatively slow, no more than 10% of the harvest area should be disturbed.

A plug is a root soil mass 10 to 15 cm in diameter and 15 to 20 cm deep which contains roots and a number of stems. Plugs can be used as an alternative to sprigs. Plugs are much more labor intensive than sprigs. However, on the Pacific coast it is often difficult to locate uncrowded stands of cordgrass from which good quality sprigs can be obtained. Consequently, plugs are used more frequently on the Pacific coast than in other regions.

Plugs are harvested from existing marshes which have heavy textured sediments (clays and silts). The vast majority of Pacific marshes are found on cohesive sediments. An intact root-soil mass cannot be maintained if plugs are excavated from noncohesive sandy sediments. Plugs should be planted slightly below the substrate surface and soil firmed tightly around them.[4,12]

Studies in San Francisco Bay have demonstrated the effectiveness of a new type of plug.[12] It has been observed for some time, that Pacific cordgrass growing in association with mussels (*Ischadium demissum* Dillwyn) form a riprap like mat which is extremely resistant to wave action.[26] Newcombe et al.[12] harvested plugs from these mats and used them to plant several eroding shores. They found these cordgrass-mussel plugs (termed "bioconstructs") to be more tolerant to wave activity than other plugs. Locating a harvestable source for this type of plug will be considerably more difficult. However, where available, bioconstructs are a promising plant material.

### c. Nursery Materials

Nursery seedlings are plants grown from seed in peat moss or plastic pots under controlled conditions. Information on seed harvest and storage is presented above in the discussion of seeding.

The seeds are removed from storage and scattered over the surface of 5 to 10 cm pots filled with sand (seeds may be germinated in flats and transplanted into pots). Approximately ten seeds are applied to the surface of each pot and covered with a thin layer of sand. The pots are then irrigated with tap water and 10-10-10 fertilizer is applied after the seeds have germinated, and as often thereafter as needed to maintain good color and growth. Slow release fertilizers may be added to the pot in lieu of the above. The growth solution should be adjusted with sodium chloride to maintain a salinity comparable to the intended planting site. Fresh water can be used for irrigation initially, but the solution should be adjusted to planting site conditions several weeks before planting. It is not necessary to maintain an artificial condition of tidal flooding in the nursery, though plants should be watered more frequently than are terristrial plants under nursery care.[4]

## B. Pickleweed (*Salicornia* spp.)

### 1. Plant Description

This plant is a frequent colonizer of the irregularly flooded portion of the intertidal zone in the more saline waters of the Pacific coast. On the southern Pacific coast, it is often the dominant species in the high marsh. On the northern Pacific it is typically mixed with other species. It is a fleshy-stemmed, weedy-type plant that spreads readily vegetatively and by seeds. It invades and covers bare areas rapidly, but is unable to persist in the regularly flooded portion of the intertidal zone. Unlike Pacific cordgrass this plant is not equipped to supply oxygen to its roots from its above ground stems. It, typically, is found in the zone between mean lower high water and the elevation of the highest tides. Pickleweed is the most salt-tolerant of all Pacific plants. It tolerates salinities more than twice sea strength, up to about 80%.[27] It is a prime candidate for marsh restoration in the irregularly flooded zone on the southern Pacific coast. On the northern Pacific and in fresher areas, pickleweed plantings should be supplemented with other species.

### 2. Cultural Techniques

Pickleweed sets seed in October and November. Seed-producing shoot tips may be collected during this period and dried. The seeds of pickleweed are very small. Stored seeds germinate readily when treated with salt water.[4] There appears to be no barriers to the direct seeding of pickleweed, though we know of no successful attempts to do so. All successful germination to date has been performed under nursery conditions.

The use of cuttings has been attempted. The plantings was not successful.[4]

Pickleweed seedlings have been produced under nursery conditions in peat-pots containing sand, vermiculite, and clay sediments. The seeds were planted dry and irrigated with salt water. The seedlings were transplanted into an intertidal environment. The planting successfully established a cover of pickleweed. However, the unplanted controls in this experiment reached about 50% of the density of planted areas after two growing seasons.[4,9] This experiment illustrates that the natural invasion of pickleweed is very rapid. Planting will be necessary only when rapid plant cover is required.

## C. Saltgrass (*Distichlis spicata*)

This grass is widely distributed along the Pacific coast. It is a colonizer of bare areas, usually in mixture with other species, in the irregularly flooded portion of the intertidal zone roughly between mean lower high water and mean higher high water. Saltgrass is seldom dominant, usually occuring in small patches or bands. Experience with this grass elsewhere has shown it to be difficult to plant but quick to naturally invade stands of other planted species. Saltgrass is second only to pickleweed in its tolerance to salinity. It withstands salinity up to about 50°/₀₀.

Saltgrass spreads vegetatively and by seeds. Transplanting success using sprigs has been poor. Survival has been low and initial growth slow. A recent report stated that success was obtained with vegetative materials of saltgrass, however, within two growing seasons the planting was displaced by invading species.[28] Successful establishment has been reported using peat pot-grown seedlings.[29] Because saltgrass readily invades established stands of other species, artificial propagation of this species is seldom warranted.

FIGURE 5.   Lyngbye's sedge. (Photo courtesy of D. L. Higley and R. L. Holton.)

## D. Lyngbye's Sedge (*Carex Lyngbyei*)

### 1. Plant Description

This plant is a major component of salt, brackish, and fresh water marshes in the Pacific northwest Figure 5). The species composition of these marshes is more diverse than on other coasts of the U. S. Less discrete elevational zones are found on the northern Pacific coast where fresh water makes a much greater impact on the distribution of species. The transition between southern Pacific and northern Pacific marshes occurs between San Francisco Bay and Humboldt Bay.[3]

Lyngbye's sedge marshes usually occur on silty substrates just above seaside arrowgrass (*Triglochin maritima*). It is found from about mean lower high water to mean higher high water. In areas where tidal range is restricted (less than 2 m) sedge may be found as low as the elevation of mean tide.[30] Sedge is less tolerant to salinity than saltgrass and pickleweed and is more likely to occur on river delta marshes. It is found in salinities from fresh water to about 20 °/oo. Successful intertidal plantings have recently been made in the Columbia River estuary under freshwater conditions.[13] There have been several other smaller plantings in brackish and saltwater.

### 2. Cultural Techniques

This plant spreads vegetatively and by seeds. Planting has been limited to sprigs gathered from the wild. This appears to be both satisfactory and practical for small

FIGURE 6.    Tufted hairgrass. (Photo courtesy of D. L. Higley and
R. L. Holton.)

to moderate size plantings. Sedge is plentiful throughout most of the northern Pacific
coast and it is easy to harvest and transplant.[13] Plants should be dug from young stands
that are less than 2 years old and each planting unit should consist of three or more
stems with attached root material. Preliminary tests using plugs have not been encouraging. It is likely that planting stock can be readily moved from high to low salinities
but not the other way around. Good quality planting stock may be readily produced
from older stands by covering them with 10 to 15 cm of sediment the year before
harvesting. As this species is abundant in the region, wild harvesting will often be more
practical than planting nurseries.

### E. Tufted Hairgrass (*Deschampsia caespitosa*)

### 1. Plant Description

Tufted hairgrass is the most prevalent plant in the highest zone of northern Pacific
marshes. Its elevational range is similar to that of sedge, from about mean lower high
water upward. However, when planted with sedge the two species should overlap at
the elevation of mean higher high water and the tufted hairgrass should be extended
somewhat higher than sedge. Hairgrass is found in freshwater and in brackish waters
up to about 20 °/₀₀. Hairgrass grows in elevated tussocks or clumps. It is a good stabilizer of sediments, it is plentiful and it is easy to plant. It is the most promising plant
for use in the upper third of the intertidal zone[13] (Figure 6).

## 2. Cultural Techniques

Experience with planting this species is limited, but it has been planted successfully. It is easy to transplant and quick to establish. Plantings have been made using material harvested from natural stands.[13] Digging and separating sprigs is relativey easy and is a reliable plant material. Hairgrass should be well suited for propagation in field nurseries because it grows readily above the elevation of normal tides. Direct seeding has not been successful.[31]

## F. Seaside Arrowgrass (*Triglochin maritima*)

Seaside arrowgrass is also plentiful in the northern Pacific coast. In most areas it will be found lower in the intertidal zone than either Lyngbye's sedge or tufted hairgrass. Seaside arrowgrass has been planted on a limited scale.[32] Multiple stemmed transplants or plugs are likely to be effective. Procedures should be similar to planting Lyngbye's sedge. When planted with sedge, the two species should overlap at mean lower high water and the arrowgrass should extend somewhat lower than sedge.

## G. Other Plants

A number of other marsh species occur on the Pacific coast; gum plant (*Grindelia* sp.), saltbush (*Atriplex* sp.), frankenia (*Frankenia grandifolia*), Jaumea (*Jaumea carrosa*), sand spurry (*Spergularia* spp.), Pacific silverweed (*Potentilla pacifica*), seaside plantain (*Plantago maritima*), spike rush (*Eleocharis* spp.), and bulrush (*Scirpus* spp.). Little is known about planting the above species. However, Morris and Newcombe[10] observed more than one half of the above species invading a planting of Pacific cordgrass during the first growing season.

## III. PLANTING

## A. Site Preparation

The width of the substrate an elevation suitable for plant establishment will determine in part the relative stability of a marsh planting. Broader marshes are more resistant to the erosive forces of waves in the intertidal environment. A practical minimum width of 6 m has been recommended for erosion control plantings.[33] In plantings sheltered from wave activity, width is not an important consideration. However, if a planting will be subject to even moderate wave activity, the 6 m planting minimum should be used as a guide. Marsh plants seldom extend below the elevation of mean tide on the southern Pacific coast or below mean lower high water on the northern Pacific. Because of these elevational constraints, the more gradual the shore slope the broader the potential planting. On steeply sloping shores, there may be little area suitable for planting. If the planting is exposed to waves and the potential planting area is not 6 m in width, the shore should be sloped or backfilled to extend it. Backfilling must be done enough in advance of planting to allow for settling and firming.

Sites exposed to more than 5 km of open water will require protection from waves during the establishment period (first growing season). Temporary wave protection structures such as sand bags, rubber tires, or stone can be used to provide temporary protection. Sites exposed to more than 10 km of open water will require more permanent shore protection devices to assure long-term stability.

Salt marsh plants must rely heavily on exposure to direct sunlight and will not grow in shaded areas. Therefore, any overstory of woody vegetation present at a site should be cleared above the planting area and landward to a distance of 3 to 5 m.

## B. Fertilization

A fertilizer response has not been demonstrated for Pacific cordgrass on fine textured substrates (clay).[4,8,9] Lyngbye's sedge was reported as very responsive to fertilization on a sandy substrate in the lower Columbia River, with little or no growth on unfertilized plots.[13] There was also some indication of fertilizer response for tufted hairgrass, but it was not as striking as on sedge.[13] It is likely that plantings on sandy sediments will benefit from applications of fertilizer. Fertilizers can also be used to encourage rapid growth in erosion control plantings.

On the Atlantic coast, both soluble and slow release fertilizers have proven to be effective in intertidal plantings.[29,34,35] Experimental evidence on this subject is much more complete for the Atlantic and it is likely that these findings are appropriate for the Pacific coast. The following general guidance on fertilization has been advanced by Knutson and Inskeep.[36]

Soluble materials should be broadcast and disced in prior to planting, spread in the planting furrow, placed in a second hole beside the planting hole, or placed in the bottom of the planting hole and covered with soil before the plant is inserted. Slow release materials such as Osmocote® or Mag Amp® should be very effective when applied in the planting hole or furrow.

If soluble materials are used, they should be applied at a rate of 100 kg/ha of nitrogen (N) and 100 kg of phosphate ($P_2O_5$) at time of planting. In conventional mixed fertilizers, such as 10-10-10, the number designations represent the percentages (by weight) of nitrogen, phosphate, and potassium, respectively, that are found in the mixture. Therefore, the amount of 10-10-10 fertilizer per hectare needed to provide 100 kg of nitrogen and phosphate would be 1000 kg. A top-dressing of an additional 100 kg/ha of soluble nitrogen (N), 6 to 8 weeks after planting, will be helpful on deficient sites and a third application 6 weeks later will be advisable on acutely deficient sites.

Slow release materials, if used in lieu of soluble fertilizer, should be applied at a rate of 100 kg/ha of nitrogen at time of planting. For conventional slow release mixtures (14-14-14 or 16-8-12), about 15 g of fertilizer should be placed in each hole. When slow release materials are used, no additional applications are necessary during the first growing season.

If plant cover and development are not adequate by the second growing season, fertilize again with 100 kg of nitrogen using a soluble source broadcast at low tide in early spring. After establishment, the color of the plant can be used as a general indicator of available nitrogen. Fertilization after the first growing season will seldom be necessary even in sandy sediments.

Tidal plants also respond to application of phosphorous. However, most conventional mixed fertilizers contain more than enough phosphorous to stimulate growth if the recommended rate of nitrogen is applied.

## C. Planting Techniques

Seeds should be broadcast at low tide and covered with 1 to 3 cm of tillage. It is usually advisable to till both before and after broadcasting to insure uniform coverage. In general, seeding should be attempted only on sites that are totally sheltered from wave activity.

The essentials of successfully transplanting vegetative materials are to

1.    Open a hole or burrow deep enough to accommodate the plant
2.    Keep the hole open until the plant can be properly inserted
3.    Insert the plant to the full depth
4.    Close the opening
5.    Firm the soil around the plant

Three to five sprigs are inserted in each planting hole. Plugs and pots are planted individually. Planting must be done during low water when the site is exposed. Planting

FIGURE 7.  Hand planting. (Photo courtesy of S. W. Broome.)

on flooded sites is inefficient and increases the likelihood of plants floating out. Hand planting, using dibbles, spades, and shovels, is the most practical method for small scale plantings, less than ½ ha (Figure 7). Normally, planting crews work in pairs, one worker opening holes and the other inserting the plant and closing the hole. Fertilizer may be dropped in the hole by a third worker during planting or fertilization may be handled as a separate operation. Machine planting of sprigs, where terrain allows, can do a much more uniform job and is far more economical than hand planting in large scale plantings. Tractor-drawn planters used for transplanting cabbage, tomatoes, tobacco, etc. are used, requiring either no alteration or simple adjustment of the row opener for certain soils. The above disk-type planters have not been used on the Pacific coast. One Pacific planting was made with a high-flotation tractor pulling a planting sled in a cohesive sediment.[9] Barriers to machine planting are inadequate traction on compact substrates, insufficient flotation on soft sites, or the presence of tree roots or other debris.

Planting depth is basically independent of the method or material used. Most species do best planted 3 to 5 cm deeper than they were growing. Where erosion is expected, even deeper planting is recommended. If, on the other hand, deposition is likely, they should be set close to the depth they were growing when dug or when removed from pots.

## D. Plant Spacing

Knutson[8] reports some success with seeding Pacific cordgrass at a rate of 100 seeds per square meter. Clairain et al.[31] report no success in seeding tufted hairgrass at the same rate. However, success in marsh seeding projects is erratic.

Test plantings in California[4,8,9] and Oregon[13,31] were successful using a spacing between plants of about 1 m. It is likely that sites that are exposed to waves should be planted at a closer spacing. A spacing of 0.5 m on exposed sites and 1.0 m on sheltered sites has been recommended by several researchers.[14-16] Planting on 1 m centers requires 10,000 planting units per hectare. One half meter spacing uses 40,000 plants per hectare.

## E. Planting Season

Morris and Newcombe[10] transplanted Pacific cordgrass at monthly intervals and concluded that survival was best for plantings made in July through December; growth was best in plantings made from April to August. Evidence is lacking on other species. To assure maximum spread during the first growing season, spring plantings (April) are preferred. However, with the exception of late winter all seasons appear to be suitable.

## F. Maintenance

Once a site is planted, it will be necessary to keep it free from debris that might smother the plants, especially during the first two growing seasons. Litter such as wood, styrofoam, algae, and dislodged marsh plants form a strandline that should be removed in the autumn and spring.

## IV. PLANTING COSTS

### A. Harvesting, Processing, and Planting

The principal cost of a project (unless site preparation and/or temporary protection is required) is the labor to obtain or produce propagules and plant them. Harvesting and planting must usually be confined to about a 4-hr period when the tide is low. This substantially effects the cost of labor. Lyngbye's sedge and tufted hairgrass sprigs have been harvested, processed, and planted by hand at a rate of about 10 man-hours per 1000 plants.[11] In San Francisco Bay, plantings of Pacific cordgrass sprigs required more than 30 man-hours per 1000 plants.[4] However, the cordgrass planting was made in cohesive sediments which greatly slowed harvesting and planting operations. In sandy sediments, Pacific cordgrass should be as efficient to plant as sedge and hairgrass. Conversely, the later plants will require more effort if they are planted in clay or silt environments. Two national studies suggest that 10 to 20 man-hours per 1000 planting units is a reasonable labor estimate for sprigs.[11,16] Planting plugs is at least three times more time consuming, 30 man-hours per 1000.[4] Preparation and planting of nursery material is comparable to plugs.[4]

To estimate labor requirements for vegetative plantings, first determine the number of planting units needed as follows:

$$\text{Number of Planting Units} = \text{Area of Planting} \times \frac{1}{(\text{Plant Spacing})^2}$$

(Plant spacing is 0.5 m for sites subject to erosion and 1.0 m for sheltered sites.)

Second, determine the labor needed to prepare and plant the required number of planting units as follows:

$$\text{Labor Required} = \text{Number of Planting Units} \times \frac{\text{Man-hours}}{1000 \text{ Planting Units}}$$

FIGURE 8.   Wave erosion of natural marsh, San Francisco Bay.

(As noted above, in sandy sediments sprigs require about 10 man-hours and plugs about 30 man-hours per 1000 units. Harvesting and planting in cohesive sediments may increase labor requirements by a factor of 3.)

A typical project using sprigs spaced 1.0 m apart will require 10,000 planting units and 100 man-hours per hectare. Actual project cost will vary greatly depending upon the local labor market, logistics and site trafficability.

Seeding is typically very efficient, about 25 man-hours per hectare. Hydromulching has been used, but with no success.[12]

## B. Fertilization

Conventional fertilizer costs are variable but will range from about $150 to $250/ha (1980 prices) the year of establishment, including labor. Slow-release fertilizer is considerably more expensive, about $1500 to $2500/ha.

## V. FACTORS CONTROLLING SUCCESS OR FAILURE

### A. Wave Stress

Commonly, the principal factor affecting initial establishment and long-term stability of salt marshes is wave stress. It is difficult to describe wave environments compatable with marsh establishment. Fetch, windspeed, wind duration, and water depth are all major determinants of wave climate. In addition, the tidal elevation associated with a particular set of waves and shore topography greatly influence the stress placed upon plantings. How the plants respond to waves will depend upon growth stage, density, vigor, and the overall width of the planting. Because of the complexity of this issue and the lack of experimental evidence, marsh plantings which are made in areas exposed to waves will be subject to an unquantifiable level of risk. Evidence of wave erosion on natural marshes can be observed on any shoreline of the larger bays on the Pacific coast (Figure 8).

The most reliable indicator of wave climate severity is fetch (the distance across open water). Knutson[15] developed planting guidelines with respect to fetch. He suggests that seeds of Pacific cordgrass should be planted only in areas exposed to fetches of less than 1.6 km. Sprigs can be planted in areas exposed to 8 km of open water and plugs up to about 16 km. Newcombe et al.[12] found these guidelines to be overly optimistic.

Woodhouse[14] estimates that vegetative material should be planted only in areas exposed to less than 5 km of open water, seeds in less than 1 km. Recently, Knutson[33] developed a method for evaluating wave climate and the likelihood of planting success on a case-by-case basis (see Chapter 2). However, only 5 of the 86 marshes surveyed to develop the method were located on the Pacific coast.

## B. Other Factors

Salinity is a major stress upon all plants growing within the intertidal zone. However, the species specified for use in this chapter are all salt tolerant. Salinity tolerance ranges for the species discussed in this chapter are presented in Sections III and VII.

Soil type will mainly effect the planting technique and need for fertilizer since most salt marsh plants have a wide tolerance range for substrates. The actual planting will be easier in loose sandy soils than in heavy plastic or very compact soils.

Plant species also have a specific preference for a particular portion of the intertidal zone. Plants should not be planted lower in the intertidal zone than they are found normally. The elevational distribution of species in this chapter is presented in Sections III and VII.

Strong tidal action can undermine plantings; therefore, location and probable migration of tidal channels in the vicinity of prospective plantings should be considered.

The presence of patches of healthy marsh on or near the site is an excellent indicator that there are no environmental factors which are likely to limit plant establishment at the site.

## VI. VALUES OF PLANTED MARSH

As noted in Chapter 2, salt marshes are valued as sources of primary production and as nursery grounds for sport and commercial fishery species.[37-41] They also stabilize and protect shorelines, store and recycle nutrients and pollutants such as nitrogen, phosphorus, and heavy metals, and form wintering areas for migratory waterfowl. Once established, marsh plantings function as natural salt marshes and gradually develop comparable animal populations.[42,43]

Research on the productivity of Pacific marshes is much less abundant than Atlantic research. However, scientists on the Pacific coast have demonstrated that primary production in these marshes is comparable to Atlantic marshes.[6,44]

Higley and Holton[7] suggest that Pacific marshes may be less important as nursery area for marine species than Atlantic marshes, though in other respects they note many similarities. Several investigators have observed relationships between salmon (*Oncorhynchus* spp.) and marsh based food chains.[45-49]

It has been demonstrated that the presence of marshes increases shore stability. First, the aerial stems of marsh plants form a mass which dissipates wave energy.[50] As wave energy is diminished, the offshore and long-shore transport of sediments is reduced. Optimally, dense stands of marsh vegetation can create a depositional environment, causing accretion rather than erosion of the shore. Second, salt marsh plants have extensive root systems which increase the stability of shore sediments.[26,51,52] These attributes have made marsh planting an attractive alternative to structural shore protection for many years.[53,54]

## VII. SUMMARY

A.    Principal Species
   1.    Pacific Cordgrass (*Spartina foliosa*)
       a.    Distribution — the California coast from Humboldt Bay south to Mexico.

        b.     Elevational range — mean tide to mean higher high water.

        c.     Salinity range — 10 to 40 $^o/_{oo}$.

  2.    Pickleweed (*Salicornia* spp.)

        a.     Distribution — throughout Pacific coast.

        b.     Elevational range — mean lower high water and above.

        c.     Salinity range — 20 to 80 $^o/_{oo}$.

  3.    Saltgrass (*Distichlis spicata*)

        a.     Distribution — throughout Pacific coast.

        b.     Elevational range — mean lower high water upward.

        c.     Salinity range — 10 to 50 $^o/_{oo}$.

  4.    Lyngbye's Sedge (*Carex Lyngbyei*)

        a.     Distribution — Northern California, Washington, and Oregon.

        b.     Elevational range — mean lower high water to mean higher high water; as low as mean tide level in restricted tidal ranges (less than 2.0 m).

        c.     Salinity range — from fresh to about 20 $^o/_{oo}$.

  5.    Tufted Hairgrass (*Deschampsia caespitosa*)

        a.     Distribution — Northern California, Oregon, and Washington.

        b.     Elevational range — mean lower high water upward.

        c.     Salinity range — near fresh to about 20 $^o/_{oo}$.

  6.    Seaside Arrowgrass (*Triglochin maritima*)

        a.     Distribution — Northern California, Oregon, and Washington.

        b.     Elevational range — mean lower high water (often lower) to mean higher high water.

        c.     Salinity range — fresh and brackish.

B.    Plant Materials

  1.    Seeds — experience is limited. If used, environment must be sheltered from waves.

  2.    Sprigs — the most common method, efficient and typically successful.

  3.    Pot-grown Seedlings — experience is limited. Awkward to transport and plant but typically successful.

  4.    Plugs — awkward to transport and plant, but an alternative to pot-grown seedlings and equally successful.

C.    Planting Methods

  1.    Hand planting (dibbles, spades, and shovels) — suitable for all plant materials.

  2.    Power-driven Auger — useful for difficult soils and for seedlings and plugs.

  3.    Machine Planting (Cabbage, Tomato and Tobacco Planters) — very efficient for large scale plantings, but only suitable for planting sprigs.

## VIII. RESEARCH NEEDS

Research and experience on planting coastal marshes on the Pacific coast is in its infancy. The guidelines presented in this chapter are based on perhaps five field planting experiments, 20 years of first hand experience by the authors, and a great deal of conjecture. Research is needed on environmental tolerances and propagation strategies for all major coastal species. This lack of experience is particularly ironic considering the major need for marsh restoration on the Pacific coast.

Research on the ecology and productivity of Pacific marshes is also lacking. A more comprehensive understanding of these marshes and their relationship to the marine environment is needed if the remaining marshes of the Pacific are to be spared and new marshes are to be constructed.

# REFERENCES

1. **Cooper, A. W.**, Salt marshes, *Coastal Ecological Systems of the United States,* Vol. 1, Odum, M. T., Copeland, B. J., and McNamon, E. F., Eds., Institute of Marine Science, University of North Carolina, Chapel Hill, 1969, 567.
2. **Barbour, M. G. and Major J.**, Eds., *Terrestrial Vegetation of California,* John Wiley & Sons, New York, 1977.
3. **MacDonald, K. B. and Barbour, M. G.**, Beach and salt marsh vegetation of the North American Pacific coast, in *Ecology of Halophytes,* Reimold, R. J. and Queen, W. H., Eds., Academic Press, New York, 1974.
4. **U.S. Army Engineer District, San Francisco**, San Francisco Bay and Estuary — Dredge Disposal Study, App. K, Marsh development, San Francisco, Calif., 1976.
5. **de la Cruz, A. A.**, The role of tidal marshes in the productivity of coastal waters, *Assoc. Southeastern Biol. Bull.,* 20(4), 147, 1973.
6. **Eilers, H. P.**, Production ecology in an Oregon coastal salt marsh, *Estuarine Coastal Mar. Sci.,* 8, 399, 1979.
7. **Higley, D. L. and Holton, R. L.**, A study of the invertebrates and fish of salt marshes in two Oregon estuaries, MR 81-5, U.S. Army, Corps of Engineers, Coastal Eng. Res. Cent., Fort Belvoir, Va., June 1981.
8. **Knutson, P. L.**, The use of dredged material for the development of intertidal marshlands, *Proc. of MTS/IEEE Ocean 75 Conf.,* San Diego, California, 1975, 208.
9. **Knutson, P. L.**, Development of intertidal marshlands upon dredged material in San Francisco Bay, *Proc. of 7th World Dredging Conf.,* San Francisco, Calif., 1976, 103.
10. **Morris, J. H., Newcombe, C. L., Huffman, R. T., and Wilson, J. S.**, Habitat development field investigations, Salt Pond No. 3 marsh development site, South San Francisco Bay, California; Summary Rep., TR D-78-57, U.S. Army, Corps of Engineers, Waterways Exp. Stn., Vicksburg, Miss., 1978.
11. **U.S. Army Waterways Experiment Station**, Wetland habitat development with dredged material: engineering and plant propagation, TR DS-78-16, U.S. Army, Corps of Engineers, Vicksburg, Miss., March, 1977.
12. **Newcombe, C. L., Morris, J. H., Knutson, P. L., and Gorbics, C. S.**, Bank erosion control with vegetation, San Francisco Bay, California, MR 79-2, U.S. Army, Corps of Engineers, Costal Eng. Res. Cent., Fort Belvoir, Va., May 1979.
13. **Tennyik, W. E.**, Pilot propagation of vascular plants on Miller Sands habitat and marsh development site, Columbia River, Oregon, 9WD D-77-2, U.S. Army, Corps of Engineers, Waterways Exp. Stn., Vicksburg, Miss., March 1977.
14. **Woodhouse, W. W., Jr.**, Building salt marshes along the coasts of the continental United States, SR 4, U.S. Army, Corps of Engineers, Coastal Eng. Res. Cent., Fort Belvoir, Va., May 1979.
15. **Knutson, P. L.**, Designing of bank erosion control with vegetation, in *Coastal Sediments '77,* American Society of Civil Engineers, New York, 1977, 716.
16. **Knutson, P. L. and Woodhouse, W. W.**, Shore stabilization with salt marsh vegetation, U.S. Army, Corps of Engineers, Coastal Eng. Res. Cent., Fort Belvoir, Va., in press.
17. **Mason, H. L.**, Marsh development study, phase one, preliminary investigation, Incl. 1, San Francisco Bay and Estuary — Dredge Disposal Study, App. K, Marsh Development, U.S. Army Engineer District, San Francisco, Calif., April 1976, 1.
18. **Harvey, H. T.**, Growth and elevational transplant studies of *Spartina foliosa* in dredged material, Inc. 2, San Francisco Bay and Estuary — Dredge Disposal Study, App. K, Marsh Development, U.S. Army Engineer District, San Francisco, Calif., April 1976, 109.

19. **Barko, J. W., Smart, R. M., Lee, C. R., Landin, M. C., Sturgis, T. C., and Gordon, R. N.,** Establishment and growth of selected freshwater and coastal marsh plants in relation to characteristics of dredge sediments, TR D-77-2, U.S. Army, Corps of Engineers, Waterways Exp. Stn., Vicksburg, Miss., 1977.

20. **Purer, E. A.,** Plant ecology of the coastal salt marshlands of San Diego County, California, *Ecol. Monogr.,* 12, 81, 1942.

21. **Floyd, K. W. and Newcombe, C. L.,** Growth of intertidal marsh plants on dredge material substrates, Incl. 3, San Francisco Bay and Estuary — Dredge Disposal Study, App. K, Marsh Development, U.S. Army, Corps of Engineers, San Francisco District, Calif., 1976.

22. **Phleger, C. F.,** Effect of salinity on growth of a salt marsh grass, *Ecology,* 52, 908, 1971.

23. **Kasapligil, B.,** A synoptic report on the morphology and ecological anatomy of *Spartina foliosa* Trin., Incl. 2, San Francisco Bay and Estuary — Dredge Disposal Study, App. K, Marsh Development, U.S. Army Engineer District, San Francisco, Calif., 1976, 112.

24. **Wong, G.,** Oxygen transport in *Spartina foliosa* Trin., Incl. 2, San Francisco Bay and Estuary — Dredge Disposal Study, App. K, Marsh Development, U.S. Army Engineer District, San Francisco, Calif., 1976, 132.

25. **Hinde, H. P.,** The vertical distribution of salt marsh phanerogams in relation to tide levels, *Ecol. Monogr.,* 24, 209, 1954.

26. **Pestrong, R.,** San Francisco Bay tidelands, *Calif. Geol.,* 25(2), 27, 1972.

27. **Mall, R. E.,** Soil-water-salt relationships of waterfowl food plants in the Suisun marsh of California, *Calif. Fish and Game Dept. Bull.,* No. 1, Sacramento, Calif., 1969.

28. **Hardisky, M. H. and Reimold, R. J.,** Buttermilk sound marsh habitat development site, Glynn County, Georgia, Georgia Department of Natural Resources, 1979.

29. **Garbisch, E. W., Jr., Woller, P. B., and McCallum, R. J.,** Salt marsh establishment and development, TM 52, U.S. Army, Corps of Engineers, Coastal Eng. Res. Cent., Fort Belvoir, Va., June 1975.

30. **Jefferson, C. A.,** *Coastal Wetlands of Oregon,* G. J. Akins, Ed., Oregon Coastal Conservation and Development Commission, Florence, Ore., August 1973, 1.

31. **Clairain, E. J., Jr., Cole, R. A., Diaz, R. J., Ford, A. W., Huffman, R. T., Hunt, L. J., and Wells, B. R.,** Habitat development field investigations, Miller sands marsh and upland habitat development site, Columbia River, Oregon, TR D-77-38, U. S. Army, Corps of Engineers, Waterways Exp. Stn., Vicksburg, Miss., December 1978.

32. **Armstrong, J.,** personal communication, 1979.

33. **Knutson, P. L., Ford, J. C., Inskeep, M. R., and Oyler, J.,** National survey of planted salt marshes (vegetative stabilization and wave stress), *Wetlands,* 1, 129, 1981.

34. **Woodhouse, W. W., Jr., Seneca, E. D., and Broome, S. W.,** Propagation of *Spartina alterniflora* for substrate stabilization and salt marsh development, TM 46, U.S. Army, Corps of Engineers, Coastal Eng. Res. Cent., Fort Belvoir, Va., August 1974.

35. **Woodhouse, W. W., Jr., Seneca, E. D., and Broome, S. W.,** Propagation and use of *Spartina alterniflora* for shoreline erosion abatement, TR 76-2, U.S. Army, Corps of Engineers, Coastal Eng. Res. Cent., Fort Belvoir, Va., August 1976.

36. **Knutson, P. L. and Inskeep, M. R.,** Shore erosion control with salt marsh vegetation, CETA, U.S. Army, Corps of Engineers, Coastal Eng. Res. Cent., Fort Belvoir, Va., 1982.

37. **Odum, E. P.,** The role of tidal marshes in estuarine production, *N.Y. State Conserv.,* 15(6), 12, 1961.

38. **Teal, J. M.,** Energy flow in the salt marsh ecosystem of Georgia, *Ecology,* 43, 614, 1962.

39. **Udell, H. F., Zarudsky, J., Dohemy, T. E., and Burkholder, P. R.,** Productivity and nutrient value of plants growing in the salt marshes of the town of Hempstead, Long Island, *Bull. Torrey Bot. Club,* 96, 42, 1969.

40. **Odum, E. P. and Smalley, A. E.,** Comparison of population energy flow of a herbivorous and a deposit-feeding invertebrate in a salt marsh ecosystem, *Proc. Natl. Acad. Sci. U.S.A.,* 45, 617, 1959.

41. **Williams, R. B. and Murdock, M. B.,** The potential importance of *Spartina alterniflora* in conveying zinc, manganese, and iron into estuarine food chains, Proc. 2nd Natl. Symp. Radioecol., USAEC Conf-670503, Ann Arbor, Mich., 1969.

42. **Cammen, L. M.,** Microinvertebrate colonization of *Spartina* marsh artificially established on dredge spoil, *Estuarine Coastal Mar. Sci.,* 4(4), 357, 1976.

43. **Cammen, L. M., Seneca, E. D., and Copeland, B. J.,** Animal colonization of salt marshes artificially established on dredge spoil, TP 76-7, U.S. Army, Corps of Engineers, Coastal Eng. Res. Cent., Fort Belvoir, Va., June 1976.

44. **McDonald, K. B.,** Coastal salt marsh, in *Terrestrial Vegetation of California,* Barbour, M. G. and Major, J., Eds., John Wiley & Sons, New York, 1977.

45. **Reimers, P. E.,** The Length of Residence of Juvenile Fall Chinook Salmon in Sixes River, Oregon, Ph.D. thesis, Oregon State University, Corvallis, 1971.

46. **Mason, J. C.,** Behavioral ecology of chum salmon fry (*Oncorhynchus keta*) in a small estuary, *J. Fish. Res. Board Can., Ottawa,* 31(1), 83, 1974.

47. **Dunford, W. E.,** Space and Food Utilization by Salmonids in Marsh Habitats of the Fraser River Estuary, M.Sc. thesis, University of British Columbia, Vancouver, Canada, 1975.

48. **Cliff, D. D. and Stockner, J. G.,** Primary and secondary components of tbe food-web of the outer Squamish River estuary, Manuscr. Rep. Ser. No. 1214, Fisheries Research Board of Canada, Pacific Environ. Inst., West Vancouver, British Columbia, January 1973.

49. **Healey, M. C.,** Detritus and juvenile salmon production in the Nanaimo Estuary: production and feeding rates of juvenile chum salmon, *J. Fish. Res. Board Can., Ottawa,* 36, 488, 1979.

50. **Wayne, C. J.,** Sea and marshgrasses: their effect on wave energy and nearshore transport, M.S. thesis, Florida State University, College of Arts and Sciences, Tallahassee, 1975.

51. **Gallagher, J. L., Plumley, F. G., and Wolf, P. L.,** Underground biomass dynamics and substrate selective properties of Atlantic coastal marsh plants, TR D-77-28, U.S. Army, Corps of Engineers, Waterways Exp. Stn., Vicksburg, Miss., December 1977.

52. **Manbeian, T.,** The Influence of Soil Moisture Suction, Cyclic Wetting and Drying, and Plant Roots on the Shear Strength of a Cohesive Soil, Ph.D. thesis, University of California, Berkeley, 1973.

53. **Phillips, W. A. and Eastman, F. D.,** Riverbank stabilization in Virginia, *J. Soil Water Conserv.,* 14, 257, 1959.

54. **Sharp, W. C. and Vaden, J.,** Ten-year report on sloping techniques used to stabilize eroding tidal river banks, *Shore Beach,* 38, 31, 1970.

Chapter 6

# LOW MARSHES, CHINA

## Chung-Hsin Chung

## TABLE OF CONTENTS

## I. INTRODUCTION

The total length of the Chinese coast is more than 32,000 km, that of the mainland coast being more than 18,000 km.[1] Tidal land has been important economically and culturally ever since history began. Most of the present coastal plains are old tidal lands. In recent years, as much as 1,940,000,000 t of river sediment all over China has been carried to river mouth and has been redistributed along plains coastline annually. This means more land resources to this country.

The author's direct experience of *Spartina* planting is mostly limited to Jiangsu and Zhejiang Provinces which account for over half of the total planting acreage, approximately 33,000 ha in this country. Consequently, more examples are drawn from these two provinces than from other provinces, although we also observed *Spartina* in Hebei, Shandong, Fujian, Guangdong, as well as Tianjin.

The chinese mu and catty are converted to hectare and kilograms, respectively, for the ease of reading.

## II. NATURAL COMMUNITY

In China, natural communities of higher plants cover a very small area of tidal land, while large expanses are barren flats. Along the coast north of the Yangtze River, salt marshes of *Suaeda ussuriensis* probably are the most important. Above this zone *Aeluropus littoralis* var. *chinensis* prevails. The common reed, *Phragmites communis*, is distributed in estuaries and small bays.

In South China there are not many extensive mangrove communities. However, the area has been reduced, year after year, due to cutting for firewood[2] and the persistence of extremely high soil acidity even as high as pH 2 after reclamation.

Some algae communities also occur on or in the tidal land.

### A. Productivity Values and Habitat Values

Productivity values data which are expressed as grams carbon per square meter (gC/m[2]) of leaf surface per time period as day or year are lacking for both salt marsh and for mangrove plants.

Habitat values of salt marshes lie in the use of reed marshes as a natural sanctuary of water fowl, such as Red Top Storks in the Estuary of Sheyang, Jiangsu Province, and many swans in Huanghua County, Hebei Province. They are less in recent years than previously. Of course, reed is commonly used as raw material for paper pulp as well as for fuel. Habitat values of mangroves are exploited as firewood, green manure, and accretion of mud for reclamation.[2] The true value of mangroves as a protection site of young fish, etc.[3] in spite of its significance, remains to be appreciated in China.

Some algae have been used as human sea food, hog feed, and for different purposes.[5,45]

### B. History of Habitat Loss

Before discussing the history of big scale loss of land due to erosion of the northern part of Jiangsu Province, it might be as well to go back still further to the history of big scale accretion as a background. In 1128 A.D., the Yellow River shifted its course southward to the Yellow Sea. This resulted in very rapid seaward advance of the mainland in history. From 1128 to 1855, the mouth of the Yellow River expanded eastward 90 km in more than 700 years, The continental area accreted amounted to 15,700 km² equivalent to one sixth of area of Jiangsu Province.[6]

In 1855, the Yellow River once again shifted northward, the source of silt having been cut off. Wave action forced the old Yellow River deltaic coast to recede rapidly. Since 1855, the 150 km coast, from Lianyungang, Xiangshui, Binhai, and northern Sheyang counties, has been the main eroding coast in China. About 1400 km² of land has been reclaimed by the sea. Within the past 50 years, average recession of sea coast has been 20 to 30 m annually and the surface of tidal land has been eroded at a rate of 5 to 10 cm/year. The former expanding coast has become a receding coast.[6]

More than 10 km of land have been lost to sea during the past 70 years in Liuhe Village, Binhai County. Still further north, Xiaodinggang, just 3 km south of Lianyungang, provides a serious case of recent destruction. In 1953, the elevation of the tidal flat was 2 m above sea level, whereas in 1976 it was 0.3 m below sea level.

## C. Documented Loss of Habitat

Although most of the mud flats in Hebei Province are expanding, prior to 1977 a small fishing village, Fengjiapu, Huanghua County, eroded so seriously that the inhabitants had to move on to two other places to live.

Even on the accreting coast (such as part of Qidong County, Jiangsu Province) repair of the earthen sea walls must be done twice a year owing to erosion. To dig, carry, pile up, and press hard the mud of tidal flat is time- and labor-consuming. Moreover, it is not a once-and-for-all solution.

In the spring tide period, people in Rudong County, just north of Qidong, are organized to watch and to repair in case a breach is found even at night.

Stone sea walls are not invincible against typhoon attacks. Falling stones, cave ins, and uneven settlement of the cement pavement, occurred in Qidong in September typhoon period, 1977.

The so-called "Tide of Congratulating Chinese New Year" scoured away 1, 2, and even 3 m of soil of earthen wall onshore in February 1979, in the two counties mentioned above.

## III. PLANT MATERIAL TYPES AND SOURCES FOR REVEGETATION

The native plant materials of salt marshes qualified for revegetation are very scarce, which is not the case for mangroves. Furthermore, they occur on high marsh and sea grasses occur in sublittoral zone. Algae are generally considered unfit for this purpose, so the author turned to big rice grass, *Spartina anglica.*

Big rice grass really consists of two species: the male sterile, the original *S. townsendii* H. & J. Groves, and the fertile species, *S. anglica* C. E. Hubbard. The male sterile species was first collected in, or just before, the year 1870 by Hill at Hythe in Southampton Waters. It is a natural hybrid of the native *S. maritima* (Curt) Fernald and the American *S. alterniflora* Loisel. The latter had been presumably brought in by shipping from some port between Boston and Newfoundland to the Itchen River close to the date 1803. It was first collected by Borrer in 1829 and recorded in 1836 from the estuary of the River Itchen where it might have been established for at least 20 years prior to its discovery.[8] Speciation of *S. townsendii* is believed to have taken place between 1860 and 1870. Its chromosome number (2n) is 62. By doubling of the chromosomes, this primary hybrid developed into a fertile amphidiploid, *S. angelica.*[9] *S. townsendii* had been used for both species prior to 1968, because there had been no name given to the fertile species.[10] *S. anglica* was chosen preferably for introduction to China on account of its better vitality, stress resistance, and versatility than those of the male sterile species.[11]

It is a perennial grass. Its height ranges from more than 10 cm in Dovey to 155 cm in Steart Flat, Bridgewater Bay, Somerset, in Britain. It varies with differences of ecological factors such as soil fertility, salinity, wave action, length of day, etc. Natural segregation prevails in this hybrid in the same habitat, each seedling being different in external morphology as well as in growth performance. There are aerial shoots as well as rhizomes which once in a while also grow upward to become an aerial shoot. There are two kinds of roots: dense, fibrous, feeding roots in the upper layers of soil and less abundant anchoring roots reaching a depth of more than a meter.[12,13]

*S. anglica* has a viviparous embryo,[13] while after-ripening exists in ungerminated seeds until mid-April in Nanking.

Secondary roots and stems developing normally originate by growth of initials in the axile of each leaf and coleoptile. In each axile, there is one central bud which may produce a stem and two other initials which may produce roots. In addition to such tillers, some buds are downwardly directed, producing rhizomes.[13]

With the production of first rhizomes, the process of tussocking begins, resulting in a circular tussock extending peripherally. Fusion with neighboring tussocks produces more or less uniform sward over the area.[8]

Staph[14] and Oliver[15] took up a program of research using *Spartina* as an agent for coastal protection and for reclamation. Oliver demonstrated the use of *Spartina* as an agent of reclamation to Dr. Lotsy whose report aroused the interest of the Dutch government.[8] The first Dutch achievement of reclaiming 490 ha of mud flats south of Sloedam, Zeeland, is the first *Spartina* polder in the world[16]

In 1929, a pamphlet on the economic possibilities of rice grass by Oliver, Bryce, and Knowles[17] was quoted by newspapers throughout the world and led to numerous requests for plant fragments and seeds. Results of trial plantings showed *Spartina* surviving and spreading in temperate countries such as Ireland, Great Britain, the Netherlands, West Germany, Denmark, Australia, New Zealand, the U.S., and China, but not in the tropical countries.[16]

Multiple use led Oliver to call it "a Gift from Heaven".[18] Certainly, it is one of the rarest plants in the plant kingdom of which so many uses can be found in a single species.

The world total acreage of these two species of *Spartina* surveyed by Ranwell[16] in the 1960s amounted to 52,400 to 68,500 acres, i.e., 21,000 to 27,000 ha.

Five batches of *S. anglica* were introduced to China for marshland creation. The Essex River Board supplied two of them and seeds were also included in one of them. Ranwell sent Poole Harbour plants and Lancashire seeds. Only one batch was from southwestern Denmark. Owing to the early propagation of the first 35 Essex plants, of which only 21 survived, their offspring now constitute the majority of *Spartina* marshes in China today.

## IV. TECHNIQUES USED IN REVEGETATION

In revegetation work the following steps must be carried out:

1.    Selection of planting material. The most frequently used method is to plant sprigs instead of starting from seeds. It is much easier and saves more time than the latter. It is best to choose vigorous shoots from the outside of actively spreading thick-stemmed *Spartina* as planting units. After digging from the soil, either washed or with soil intact, they are put in plastic bags if the number of plants is small, and are put in sacks or other vessels if the number of plants is large.

2. Transportation of planting materials. Air transport is used for shipments from Britain to Canada and then to China. Express passenger trains are used on land, while freight trains are not desirable. Trucks and ships also serve the purpose in case there is no more rapid transit, provided roots are kept moist by sprinkling with water at all times. A batch of *Spartina* from Sheyang, Jiangsu shipped to Tianjin in June to July 1965, was in healthy condition upon arrival and produced a good planting. However, poor survival or complete failure of planting occurred after 2 weeks in a freight car without any watering on the way to an inland region in July of the same year.

3. It is important to make a careful and detailed study of the proposed planting sites and associated ecological factors, especially tidal currents, wave characteristics, wind, exposure, soil, etc.

4. Planting — on mud flats with protected physiography, such as bays, harbors, etc. 1 to 3 or 3 to 5 sprigs preferably are used with soil. Sprigs should be set so that the lower portion of shoots and feeding roots are just below the mud surface. After this, mud must be filled in and pressed hard to prevent its flowing out by wave and tide. This is very important. On hard substrates as silt and sand, a hole must be dug with a shovel or dibble first before setting the sprigs. Filling and pressing the soil to the hole are absolutely necessary to ensure success. In exposed sites, more sprigs are used in a single planting and depth of planting must be greater and spacing must be closer than in sheltered sites.

Preliminary, small scale planting[20] is a prerequisite to large scale plantings. It usually takes about a year to observe growth performance to determine the best elevation and the most favorable time of planting (Figures 1 and 2). In protected situations, probably less than a year is sufficient, while in more exposed stormy coasts probably 1 year is not enough. Large scale plantings first follow the determined best elevation of planting which is clearly indicated by the widest diameter of tussocks. This forms a tidal flat nursery which supplies planting materials upward and downward. This saves a lot of labor as well as time.

Planting times do vary from place to place in China. It is preferred to choose those months without storms, gales, or typhoons. In other words, long, calm periods facilitate success. In South China, from December to April there are no typhoons, so this is the best planting time. In Zhejiang Province (Wenling County) planting during the first half of the year is better than in the second half. April, May, and June seem to be the best planting time.[20] Due to the excessive drought in July and August, it is very difficult for the plant to survive. However, in North China, for example, May is too windy and June is the best planting time.[21] Winter planting is permissible in places where soil is not frozen since more manpower is available then.

## V. FACTORS CONTROLLING SUCCESS AND FAILURE OF REVEGETATION

### A. Physiography of Planting Sites

Generally, success is achieved in sheltered sites, while difficulties are encountered in exposed sites, with quite a number of failures. The open coast, if expanding, also may result in success. In China, transportation of silt is very heavy. In the case of the Yangtze, 485,000,000 t of sediment are carried to sea annually. Natural accretion is promoted by this enormous amount of sediment. That is why *Spartina* marsh can be created on some of the exposed tidal flats. Just to give some idea of its influence to accretion, tidal flats expand an average of 100 m seaward annually in Dafeng and Sheyang counties, Jiangsu. Since the Yellow River transports more silt than any other river in the world, the figure available is 1,145,000,000 metric tons annually.

FIGURE 1. Mud flat in Wenling planted with *Spartina*; planted in May 1965.

FIGURE 2. Sward formed 1½ year with 1 m spacing and 2½ years with 2 m spacing. (Photo taken December 1967.)

## B. Tide

Tidal submergence twice daily satisfies the water requirement by *Spartina*, which normally lives in saturated soil. It does well in submergence of 6 hr twice a day. If it is under water too long, its photosynthetic period is less than normal and too short of a submergence means too long of an exposure to drought. New growth of roots after transplanting usually takes about 2 weeks or so in Wenling. If newly planted sprigs are washed out by tide, new root growth is impossible. So this is the critical stage in determining success or failure.

## C. Waves

Waves, tides, and winds interact intimately. Waves are probably one of the most destructive agents to revegetation on intertidal zones. They wash away not only the recently planted sets, but also tussocks several months old. Our large tussocks with a diameter of 60 to 70 cm were all wiped out a short time after planting in Chuansha, east of Shanghai.

Big waves are always generated by winds. A physiography of very broken nature appeared after impacts of winds and waves to a silt flat in Chuansha. After a time, the original smooth surface was restored. Changes took place back and forth several times in a single year. The following observations were noted. At first a depression was formed around tussocks with soil scoured, then damages to plants include: (1) decrease of shoot number, (2) rhizome growth prevented, (3) abnormality of green coloration, (4) stunted growth even in large tussocks, (5) strong sprigs degraded to weak ones, (6) decimation of tussocks and individual plants by serious mud accretion, and (7) tussocks partially turned over with roots exposed.

In the first trial planting in 1966 in Chuansha, 77% of planted *Spartina* sprigs were floated out and lost by waves generated by the northeast wind coinciding with high tide and by long submergence with water masses stacked in higher parts of the flat.

## D. Wind

Stormy winds, gales, and typhoons are all very destructive to plantings. More than 200 tussocks were washed away during an August typhoon, 1967, in Chuansha. South China suffers most.

## E. Drought

Drought and concurrent high temperature cause failures of planting. In Hepu County, Guangxi, a summer drought was responsible for the failure of nursery plants in 1977. In 1978, a drought virtually transformed a mid-tidal zone into an upper tidal zone regarding frequency of tidal submergence in Dafeng County, Jiangsu. As a result of this, both survival and tillering were low. In Hebei, very long periods of drought in recent years has caused poor growth.

## F. Sudden Temperature Fluctuations

In recent years, great fluctuations of temperatures within 24 hr occurred in Hebei Province. *Spartina* plants were frozen to death.

## G. Ice Carrying Away *Spartina*

This happens very often during a thaw. The ice flows away with tussocks and soil attached. About one third of the plantings of 1965 in Tianjin were lost. The cutting of grass shoots before ice formation has been a remedy. As soon as clumps coalesce to form a sward, the ice is unable to do the same damage as to the newly planted plants.

## H. Super Accretion

This is probably a very rare phenomenon in most parts of the world. In August 1966, a sudden accretion of meter deep silt entirely stifled our *Spartina* plantation of 0.2 ha on a river bar of Chien Tang. Still much larger, 13.3 ha of *Spartina* established by the First Farm, Pingyang County, Zheziang, suffered the same after 3 years of healthy growth.

## I. Human Activities

Trampling very easily puts an end to young life on tidal flats. Digging *Nereids* in the root region causes uprooting and flowing away of many *Spartina* plants. Damages due to other forms of human activities are very harmful in the early stages of growth.

## J. Animals

The Chinese wild geese, a species of migratory birds, ate all the tussocks of 2.3 ha planted in Hetao Commune, Laoshan County, near Qingdao City, in 1967. Even rhizomes were plucked up because they tasted sweet. As soon as a sward is formed however, it is impossible to repeat the wholesale destruction.

## VI. HABITAT VALUES OF CREATED PLANT COMMUNITY

### A. Nesting and Feeding Grounds of Migratory Birds, Water Fowl, and Domestic Fowl

The early discernible value of this created plant community is exploitation by migratory birds and water fowl as nesting and feeding grounds. Since appearance of this new community in Wenling, the wild geese have no more interest to visit their old feeding place, a winter wheat field.

Recently the domestic gray geese have been fed with fresh and palatable *Spartina* in Yühuan County, southeast of Wenling. It only takes 70 days of feeding in order to be sold for export.

A study of geese feeding by Shandong Technical School of Animal Husbandry and Veterinary Medicine showed 175.8 and 470.4 geese fed with minimum and maximum yields of *Spartina* per hectare, respectively.

Haiyang County, Shandong, experimented with 1000 geese fed with *Spartina* for 2 months. Each goose increased its body weight from 1 to 2.5 to 3.0 kg.

### B. Animal Fodder

According to reports from different coastal regions of China, mule, cattle, horse, water buffalo, sheep, goat, milk goat, rabbit, and hog are all fond of eating *Spartina.*

In Dongfang Commune, Wenling, beginning in 1969, 350,000 kg yearly of fresh *Spartina* grass was cut for stock feed. The animal husbandry industry in this area has been greatly developed. Shinqi Commune, Chenghai County, Zhejiang, cut 150,000 kg of Spartina in 1972 to fatten the sheep.

A potential assessment of habitat value translated to monetary terms was tried by Yangzi Commune, Wei County, Shandong. The total area of tidal flats appropriate for growing *Spartina* was calculated to be approximately 66,667 ha. Assuming annual grass production reaches 1250 kg/ha, and the average consumption of one head of beef cattle is 2500 kg, then 30,000 head of beef cattle could be fed. If average beef production is 400 kg per head and 1 kg of beef sold at 2 Yuan, then the potential gain would be 24,000,000 Yuan.

### C. Marshland Pasture

The first *Spartina* marsh used as a grazing pasture in China began in May 1976, in Qidong, Jiangsu. Its carrying capacity, including hay cut for winter feed, is very high, 0.2 to 0.27 ha of marshland being sufficient to raise one sheep or one goat. There were 60 sheep and goats grazing in 1976 and 2300 in spring, 1981 (Figure 3).

### D. Aquaculture

The first trial was made in a commune in Qidong in 1978. The fresh *Spartina* was very readily consumed without any leftovers; 1200 kg of grass was cut and transported to the fishponds daily. Unfortunately, this experiment was discontinued in 1979.

Shortage of fish feed is getting more acute and makes collection and transportation from far away necessary. Planting grasses such as Sudan grass as fish feed to increase production has been tried, but this measure demands more land for cultivation. To plant *Spartina* around the fishpond marginland would be of some help in solving this difficult problem.

### E. Green Manure

Since 1968, Wenling commune members have cut Spartina to apply as green manure to different crops such as rice, sugarcane, etc. Since the last crop of rice badly needs green manure, the value of *Spartina* is obvious. Yield increases of late rice from 1968 to 1979 were estimated, based upon total *Spartina* harvested and the average rate of

FIGURE 3.  The first marshland pasture in China, Qidong, Jiangsu.

yield increases, to be 7.7 to 20 million kg. Commune members initiated a series of experiments, showing a 562.5 to 677.3 kg/ha of rice increase in yield in coastal communes, whereas 750 to 1125 kg/ha increases were seen in noncoastal communes.[22]

Henghe Commune cut 3 million kg *Spartina* just prior to empoldering in 1972. Average per hectare increase in yield of rice was 18%, some production brigades even increased their per hectare yield by 23%.

It has been found that 50 kg of *Spartina* is equivalent to the fertilizing effect of 0.5 kg of urea (carbamide). Inland commune members prefer to exchange their urea for *Spartina* on account of its much longer effect than chemical fertilizers. Furthermore, soil structure is greatly improved by its application. It also proved to be superior to hog stable manure.[22]

Trials in other places give similar positive results. A 900 kg/ha increase in yield of rice was reported by Lianyungang, Jiangsu.[23]

The pioneer work in cotton by Shinqi Commune, Chenghai County, Zhejiang, showed a 28.98% increase in yield over controls.

## F. Amelioration of Saline Soil

*Spartina* marsh improves structure of saline soil and adds an enormous amount of organic matter to it. Following the regular course of plant succession from bare tidal flat to stages of salt marsh communities, it requires from 10 to more than 20 years to accumulate the same amount of organic matter accomplished within a few years by planted *Spartina*. The common reed (*Phragmites*) spreads into *Spartina* swards readily in Dafeng County, Jiangsu. However, they do not survive after planting on seaward barren flats.

A striking case may be cited to demonstrate how fertile *Spartina* marsh soil is. Without any manure or fertilizer applied to a 4-year old reclaimed marshland, two consecutive rice crops totaled 8587.5 kg/ha.[24]

The following two experiments of amelioration of saline soil after empoldering by means of *Spartina* were successful. From May 1973 to May 1975, *Spartina* was planted in one plot and Indian Vigna bean in 1973 and *Sesbania* in 1974 were planted in control plot. Oil rape seeds were sown in late October 1974. Soil salinity decreased to 0.3% in the former and to 0.5% in control. There were 262.5 kg/ha of oil rape seeds harvested, a 76% gain over that of control.[25]

FIGURE 4.    Earthen sea wall behind *Spartina* marsh and stone sea
wall behind the mud flat in Chenghai, Zhejiang. A front view. (Photo
taken in September 1973.)

Rudong County, Jiangsu started in May 1979 to grow *Spartina* and *Azolla* in paddies
which were also used as fishponds. After drainage in autumn, 1980, a 12 HP tractor
could not break even a bit of soil due to interlocking of huge amounts of root systems.
A 35 HP tractor barely managed to do it and a 50 HP tractor did all the plowing.
Transplantation of oil rape seedlings were successful; *Sesbania* and barley did well
also. On the other hand, neighboring saline soil crusts formed as excessive evaporation
occurred. Since 1949, 16,000 ha of tidal flats have been reclaimed in this county, yet
only 4000 ha have been under cultivation. Evidently, this situation will be changed
from now on.

### G. Stabilization of Coast

Even an expanding coast is liable to be eroded. In 6 years, in a *Spartina* marsh near
the mouth of Shinyang River, Sheyang County, Jiangsu, 2.4 m of recession was observed. By sharp contrast, 7.2 m of recession was measured in adjacent tidal flat.

A case of enhanced resistance of earthen sea wall with protection by a created *Spartina* marsh was observed in 1971, in Chenghai County, Zhejiang. The adjacent earthen
sea wall was broken through with only a barren flat in front of it. Replacement by a
stone structure costed them 4200 Yuan for purchase of rocks and 20,000 man-days in
labor. Nevertheless, scour was not completely eliminated. The creation of a *Spartina*
marsh was chosen to assist with protection of the wall (Figures 4 and 5).[26,27]

A small trial experiment of setting *Spartina* sod blocks on an earthen sea wall was
performed in Qidong. The area was only 380 m² and the length was merely 20 m. Two
weeks after completion of the sod block setting, a typhoon with a velocity of 28 m/
sec eroded on adjacent portion of sea wall slope to a depth of 20 to 40 cm and 20 to
70 cm back. With the exception of a very small part, which was not as firm and compact (from poor quality of work), *Spartina* sod remained intact and looked as healthy
as if just set.[28] The local populace was very impressed by this splendid performance.
By the end of 1980, it had been extended to 1400 m in length.

FIGURE 5.    Same as Figure 4, side view.

## H. Reclamation of *Spartina* Marsh

Based upon our experience in establishing a small *Spartina* marsh of 0.87 ha and preliminary trials to acquire fundamental knowledge of its ecology, a large *Spartina* marsh of approximately 266.7 ha was created in 3 years, 1966 to 1968 (Figures 6 and 7). From then on to 1973, mud accretion of 80 cm was recorded. In 1974, empoldering of this marsh and flat of upper tidal zone was completed. Up to the end of 1980, total income drawn from agricultural produce amounted to 2,500,000 Yuan, more than double of the empoldering expenditure (960,000 Yuan excluding labor cost) (Figure 8).

## VII. RECOMMENDED TECHNIQUES

### A. Propagation

#### 1. Propagation from Seeds

First of all, the fungi and molds on seed coat must be washed away with distilled water before disinfection. Then seeds are put on a sheet of filter paper in a petri dish. Some seeds germinate on saturated filter paper as well as in refrigerator stored moist, while others do not. As difficulties arise in water culture of seedlings, either due to exuberant growth of algae or due to unbalanced nutrient solution, it is best to proceed with soil culture. An earthen glazed vessel with a diameter of 70 cm was used after full growth of seedlings in pots of 10 cm in diameter. Seedlings did not do well until holes in bottom of pots were filled and blocked by cement.

It took us about 3½ months, from mid-February in greenhouse to early June (in the open) to the stage of formation of five to six leaves. This was the turning point marking the beginning of tillering. Starting from 44 seedlings, 30,601 sprigs were obtained on April 26, 1965, the next spring. The multiplication rate was about 695 times. Very fertile pond mud was employed and the soil was always kept moist. From the beginning of November on, they were grown in the greenhouse.

#### 2. Propagation from Sprigs

This method is the most frequently used. The procedure is the same as above with the omission of seed germination. We started with six sprigs in May 1964, in one

FIGURE 6.   Three months after setting of sprigs in creating large
scale *Spartina* plantation. (Photo taken in December 1966.)

FIGURE 7.   *Spartina* sward formed after 1½ year growth. (Photo
taken December 1967.)

earthen vessel. After 11 months and 11 days 3607 sprigs were propagated with a multiplication rate of 601 times. Eight earthen vessels were used instead of only one vessel
at the very beginning.

### 3. Propagation from Rhizome

This method is very rarely employed except in urgent need. We found this unusual
way of propagation by trial and error. A number of leftover rhizomes after shoot

planting were set into pot soil. After 4 days, roots began to sprout and shoots later arose from one of them. After pot propagation for 375 days, they were transplanted to a paddy of 0.013 ha. As soon as sprigs coalesced, they were once again transplanted to paddies of 0.3 ha. From beginning to end it took 2 years and 4 months to propagate about 9,100,000 sprigs, the first batch transplanted on 40 ha out of 266.7 ha of *Spartina* created marsh, the largest in China in the 1960s.

### 4. Propagation in Paddies

Depending on the date of transplantation on tidal flats, the spacing of 30 cm suits early planting dates and the spacing of 50 cm suits later dates better. Regular practices of rice paddies are required such as irrigation, application of manure and fertilizers, pest and weed control, etc. Our nursery paddies were 1.3 ha in size. Plants propagated from the first crop were more than sufficient to transplant 140 ha of tidal flat with meter spacing in our experiment, but also met the needs of Dongfang Commune.

It is of utmost importance to select nursery paddies without any water leaching. Otherwise, a lot of trouble surely would defeat the purpose. A shallow layer of water should be maintained or at least the soil must be kept moist all the time. As soon as sprigs began to meet each other, it was considered to be the best time for transplantation. Otherwise, the sprigs grow old and become unsuitable planting materials.

Fertilization experiments showed that the number of propagated sprigs were usually double that of the control.

### B. Stress or Acclimation Experiments

At the same time, stress experiments are to be recommended to those places with ecological conditions very different from those of the natural habitat of the tried plant. For instance, the average temperature of the hottest month of the year, August, is 16°C in Essex, England, while that of Sheyang (its first settlement) is 26.6° C and that of Xiaoshan, south of Hangzhou (its second settlement southward), is 28°C. Our experiments were carried out in direct sunlight instead of under a mat, screening off solar radiation in the first summer in China. Actual air temperature in *Spartina* tussocks were measured. To our surprise, *Spartina* not only survived 40.5 to 42.0°C for 8 days straight, but tillering also continued.

Cold stress testing in Nanjing was also encouraging — a 23% increase of sprigs through tillering and less through rhizoming from December 1964 to February 1965.

Other stress experiments were performed in preliminary trials as well as with small scale trials. Sometimes large scale plantings yield reliable information on certain unsolved problems such as whether *Spartina* tussocks could cope with very rapid accretion or not.

### C. Transplanting Techniques

In addition to those mentioned in Section IV, the following points are recommended:

1. The most favorable time for planting in a tidal cycle is a period of several days in neap, gradually turning toward spring so sprigs just set are guaranteed to be submerged by increasing tide day by day. If planting time is in a period of several days after spring turning toward neap, then sprigs set are not submerged by decreasing tide day by day and are exposed to the danger of increasing drought.
2. Weather forecast, especially wind, wave, etc. must be taken seriously in order to avoid loss or failure.
3. To study weather and tidal data carefully and thoroughly before action is taken, draw a curve of every day tidal submergence to the elevation on a chart and this curve indicates the best elevation of planting.

FIGURE 8.    Barley field of Union Polder which is the reclaimed *Spartina* marsh. (Photo taken in 1978.)

4.    Since the *Spartina* zone is below that of upper marsh plants such as *Suaeda* and *Zoysia*, the lowest margin of those plants may be used as an indicator line of uppermost planting of *Spartina*.

5.    If sprigs are washed away or turned over by waves, they should be replanted as soon as possible.

6.    Locally propagated sprigs are preferred on account of better survival and performance than those transported from other places. So it is best to establish a local nursery either on intertidal zone or in paddies. Earthen vessels also serve the purpose.

## VIII. RESEARCH NEEDS

1.    Pure and better stocks ought to be selected and used in further plantings to improve the performance and productivity. Other useful marsh plants are to be introduced and tried. *Spartina* height is shorter in China than in Britain, in some places much shorter.

2.    There is a need to solve the difficult problem of establishing *Spartina* sward on eroded flats which badly need the protection of plants. Schleswig-Holstein brushwood dams set in regular pattern over wide mud flats have been introduced since 1900. By such means, the Germans were successful in substantially ameliorating erosion. In the 1930s, the Dutch made more improvements and *Spartina* plants were set in the system of brushwood groynes.[29]

3.    Larger scale establishment of *Spartina* marshes in several counties near Tianjin Harbor, where a serious situation of waterways occurs, should be tried. The same problem exists with other rivers in North China, to what extent the establishment of *Spartina* marshes can improve the situation remains to be seen.

4.    The suffocation of some shell-fish in which clams have been reported by people in Qidong and some other counties of Jiangsu needs to be studied. This phenomenon was mentioned only very briefly in English literature.[30]

5.    To study nutrient recycling to elucidate whether *Spartina* decreases the nutrient pool without return or some replenishment takes place in the near sea.

# REFERENCES

1. **Wang, Y.**, The coast of China, *Geosci. Can.,* 7, 109, 1980.
2. **Zhang, H. D., Zhang, C. C., Wang, B. S., and Wu, H. M.,** *Vegetation of Leizhou Peninsula,* Science Press, Peking, 1957, 69.
3. **Heald, E. and Odum, W. E.,** *The Contribution of Mangrove Swamps to Florida Fisheries,* Proc. Gulf, Caribb. Fish. Inst., 22nd Annu. Session, Miami, Fla., 1970, 130.
4. **Li, L. C.,** A general view of the Chinese algae and their economic value, *J. Chin. Bot.,* 1, 119, 1934.
5. **Li, L. C.,** An overview of investigations on the Chinese algae and their economic importance, *Science,* 26, 94, 1943.
6. **Zhu, D. K.,** On the Problem of Coastal Development of Central Jiangsu, unpublished, 1981.
7. **Lianyungang City Vegetable Crop Experimental Station,** A Report of Sea Defense Effect of *Spartina angelica* marsh, 1977, 1.
8. **Hubbard, J. C. E.,** *Spartina* marshes in Southern England, VI. Pattern of invasion in Poole Harbour, *J. Ecol.,* 53, 799, 1965.
9. **Marchant, C. J.,** Evolution in *Spartina* (Gramineae), I. The history and morphology of the genus in Britain, *J. Linn, Soc. (Bot).,* 60, 1, 1967.
10. **Hubbard, C. E.,** *Grasses,* 2nd ed., Penguin Press, London, 1968.
11. **Bryce, J.,** personal communication, 1963.
12. **Goodman, P. J., Braybrooks, E. M., and Lambert, J. M.,** Investigations into 'die-back' in *Spartina townsendii* agg, I. The present status of *Spartina townsendii* in Britain, *J. Ecol.,* 47, 651, 1959.
13. **Goodman, P. J., Braybrooks, E. M., Marchant, C. J., and Lambert, J. M.,** 4. *Spartina townsendii* H. & J. Groves sensu lato in biological flora of the British Isles, *J. Ecol.,* 57, 298, 1969.
14. **Stapf, O.,** Townsend's grass or rice grass, *Proc. Bournmouth Nat. Sci. Soc.,* 5, 76, 1914.
15. **Carey, A. E. and Oliver, F. W.,** *Tidal Land, A Study of Shore Problems,* Blackie & Son, London, 1918, 176.
16. **Ranwell, D. S.,** World resources of *Spartina* marshland, *J. Appl. Ecol.,* 4, 239, 1967.
17. **Oliver, F. W., Bryce, J., and Knowles, F.,** *Economic Possibilities of Rice Grass,* Misc. Publs. Minist. Agric. Fish, London, 1929, 66.
18. **Oliver, F. W.,** Blakeney Point Reports, *Trans. Norfolk Norwich Nat. Soc.,* 12, 630, 1929.
19. **Ranwell, D. S.,** personal communication, 1963.
20. University of Nanking *Spartina* Research Group, *A Brief Report of Spartina Planting Experiments on Dongpian Tidal Flat, Wenling County, Zhejiang, Amelioration and Use of Saline Soil,* No. 4, Special Number on *Spartina,* Zhoushan Prefacture Office of Science and Technology, Zhejiang, 1975, 10.
21. Hebei Provincial Estuarine Experimental Station, *A Report of Spartina anglica Introduction Experiments,* 1981, 1.
22. Shinhe District Agricultural Extension Service, Wenling County Office of Agriculture and Forestry, Zhejiang, *News Lett. Agric. Sci. Technol.,* 6, 24, 1974.
23. Lianyungang City Vegetable Experimental Station, An experiment of *Spartina anglica* fertilizing effect to rice, *Lianyungang Sci. Technol.,* 2, 15, 1977.
24. Taizhou Prefacture Institute of Agricultural Science and Wenling County August First Polder Experimental Station, Productivity of *Spartina* marsh and experiments of methods of fertilization, *Wenling Sci. Technol.,* 3, 6, 1979.
25. Taizhou Prefacture Institute of Agricultural Science, *A Preliminary Report of Amelioration of Empoldered Tidal Flat and Its Effect to Crop Yield, Amelioration of Saline Soil and its Use,* No. 2, Zhejiang Provincial Office of Science and Technology, 1976, 693.
26. University of Nanking *Spartina* Research Group, A Vanguard to Occupy the Tidal Flats, Big Rice Grass, *Sci. Exp.,* 9, 13, 1974.
27. University of Nanking Institute of *Spartina* and Tidal Land Development, A Pioneer Plant of Exploitation of the Tidal Land, *Spartina angelica,* 1980, 2.
28. **Chung, C. H., Zhuo, Y. Z., and Tsao, H.,** Big Rice Grass, a Pioneer Plant of Exploring Tidal Flats, *Nantung Prefact. Sci. Technol.,* 3, 15, 1978.
29. **Bryce, J.,** The economic exploitation of rice grass, *Emp. J. Exp. Agric.,* 9, 167, 1941.
30. **Cole, G.,** The Use of Certain Plants as Stabilizers of Marine Sediments, *J. Inst. Water Eng.,* 14, 445, 1960.

Chapter 7

# LOW MARSHES, PENINSULAR FLORIDA

## Roy R. Lewis, III

## TABLE OF CONTENTS

## I. INTRODUCTION

Previous chapters in this volume have dealt in detail with the restoration and creation of tidal marshes along the Atlantic coast (Knutson and Woodhouse), the western Gulf of Mexico (Webb), and the Northeast Gulf of Mexico including Florida (Kruczynski).

This chapter will not repeat any of that information except to refer the reader to those chapters and the general literature on southeastern U.S. tidal marshes.[1-6] This chapter will be limited to discussing the usefulness of the smooth cordgrass *Spartina alterniflora* Loisel in situations where it is not the dominant member of the coastal plant community.

## II. THE NATURAL PLANT COMMUNITY

The southern half of Florida has, as its dominant coastal estuarine vegetation, mangrove forests composed of *Rhizophora mangle* L., *Avicennia germinans* (L.) Stearn, and *Laguncularia racemosa* (L.) Gaertn. f.[7-8] A less conspicuous member of this community is smooth cordgrass (*S. alterniflora* Loisel.) which typically forms a band in front of the forest (Figure 1) in deeper water than can apparently be colonized by mangrove propagules. Davis[9] has noted that "*Spartina alterniflora* grows best in the outer deep-water zone of swamps and may be effective in holding soil materials, thus aiding in establishment of a pioneer *Rhizophora* family". He also cites Fryberg[10] as recognizing "that the mangrove plants contribute to the accretion processes, from the pioneer stage of *Spartina brasiliensis* on the offshore mud flats to the tall, close stands of red mangroves of the swamps".

## III. THE ROLE OF SMOOTH CORDGRASS IN THE RESTORATION AND CREATION OF SOUTH FLORIDA COASTAL WETLANDS

The normal practice when attempting to restore or create a coastal wetland in South Florida has been to attempt to replant mangroves since they are the dominant "climax" community (see Chapter 8).

Attempts have been made, however, to utilize *S. alterniflora* in place of mangroves[11-16] for several reasons:

1.    More rapid coverage and stabilization of substrates[11-16]
2.    Establish preferred nesting habitat for Clapper Rails (*Rallus longirostris*) and Willets (*Catoptrophorus semipalmatus*)[16]
4.    More cost effective[16]
5.    Greater value in erosion control[15]
6.    Greater value in trapping sediments in street runoff[14]

In addition, as discussed in the mangrove chapter of this volume, *S. alterniflora* acts as a nurse species trapping floating propagules of mangroves and eventually being replaced by mangroves as they shade out the grasses.

Martyn[17] in reporting on the foreshore vegetation of Georgetown, British Guiana noted: "The part played by *S. brasiliensis* in the primary colonization of the newly raised mud bank is of interest... the appearance of a grass in this role would appear to be unusual in the tropics, where muddy foreshores are more usually colonized directly by "mangrove"... once established, these plants of the mangrove association climax grow very quickly, becoming dominant to the *Spartina*, which finally almost entirely disappears beneath them."

Ranwell,[18] in discussing successional processes in *S. anglica* marshes in Great Britain

FIGURE 1. Typical view of a stand of smooth cordgrass (*Spartina alterniflora* Loisel.) situated in the lower intertidal zone in front of black mangroves (*Avicennia germinans* (L.) L. Tampa Bay, Fla.

comments on a similar shading out of *S. anglica*: "It was found that *Spartina* retained dominance for about 20 years, but in the subsequent 12 years, about 50 percent of the *Spartina* had been replaced by the invading species..." These invading species came in along the landward edge of the marsh and included *Agropyron pungens, Scirpus maritimus,* and *Phragmites communis.*

   *Spartina* also lends itself well to wetlands *creation* where upland areas may be excavated to tidal elevations and planted for mitgation.[19] Figures 2, 3 and 4 show the planting and growth of a 1.8 ha tidal marsh that was excavated from uplands and planted with 2127 plugs of *S. alterniflora* removed from an adjacent marsh. Mangroves, *Avicennia germinans* in particular have now begun invading the marsh.

   Figure 5 shows aerial views of a man-made dredged material island in Tampa Bay, Fla., where natural colonization by *S. alterniflora* was observed.[11] Figures 6 and 7 show how the colonizing plants eventually spread and coalesced and are now being replaced by both *A. germinans* and *L. racemosa.*

   *S. alterniflora* has been used specifically for its sediment binding qualities by planting it on dredged material in Florida.[11-12,16] It has the advantage of stabilizing these shifting sand substrates much more quickly than planted mangroves and can do the job for ½ to ⅓ of the cost of mangrove tree transplanting.[16]

## IV. NEEDED RESEARCH

   Development of strains of *S. aterniflora* for their genetically superior abilities in rate of growth and eventually stand height and robustness are needed. Good seed sources are also needed since most seed harvested in Florida to date is either sterile or damaged by insects.

FIGURE 2.    View of a 1.8 ha marsh creation site along Archie
Creek, Tampa, Fla., April 1978.

FIGURE 3.    Archie Creek marsh site, August 1978.

FIGURE 4.    Archie Creek marsh site, April 1979.

FIGURE 5. Vertical aerial photographs of a dredged material island in Tampa Bay, Fla.

FIGURE 6. Ground level view of a dredged material island in Tampa Bay, Fla.

FIGURE 7. Ground level view of a dredged material island in Tampa Bay, Fla.

# REFERENCES

1. Kurz, H. and Wagner, K., Tidal marshes of the Gulf and Atlantic coasts of northern Florida and Charleston, South Carolina, Florida State University, Tallahassee, Studies No. 24, 1957.
2. Adams, D. A., Factors influencing vascular plant zonation in North Carolina salt marshes, *Ecology*, 44, 445, 1963.
3. Eleuterius, L. N., The marshes of Mississippi, *Castanea*, 37, 153, 1972.
4. Eleuterius, L. N., The distribution of *Juncus roemerianus* in salt marshes of North America, *Chesapeake Sci.*, 17, 289, 1976.
5. Humm, H. J., III., The biological environment, a salt marshes, in *A Summary of Knowledge of the Eastern Gulf of Mexico*, Jones, J. I., Ring, R. E., Rinkel, M. O., and Smith, R. E., Eds., The State University System of Florida, Institute of Oceanography, St. Petersburg, 1973, III A-1.
6. Eleuterius, L. N. and McDaniel, S., The salt marsh flora of Mississippi, *Castanea*, 43, 86, 1978.
7. McNulty, J. K., Lindall, W. N., and Sykes, J. E., Cooperative Gulf of Mexico estuarine inventory and study, Florida: Phase 1, area description, NOAA Tech. Rep, NMFS Circ. 368, 1972.
8. Detweiler, T., Dunstan, F. M., Lewis, R. R., and Fehring, W. K., Patterns of secondary succession in a mangrove community, Tampa Bay, Florida, in *Proc. 2nd Annu. Conf. Restor. Coastal Vegetation Fla.*, May 17, 1975, Lewis, R. R., Ed., Hillsborough Community College, Tampa, Fla., 1976, 52.
9. Davis, J. H., The ecology and geologic role of mangroves in Florida. Paper from the Tortugas Lab., Vol. 32, *Carnegie Inst. Washington Publ.*, 517, 303, 1940.
10. Fryberg, B., Zerstorung und Sedimentation an der Mangrove-kunste Brasiliens, *Leopoldina*, 6, 69, 1930; as cited by Davis, 1940.
11. Lewis, R. R. and Dunstan, F. M., The possible role of *Spartina alterniflora* Loisel in establishment of mangroves in Florida, in *Proc. 2nd Annu. Conf. Restor. Coastal Vegetation Fla.*, May 17, 1975, Lewis, R. R., Ed., Hillsborough Community College, Tampa, Fla., 1976, 82.
12. Lewis, R. R. and Lewis, C. S., Tidal marsh creation on dredged material in Tampa Bay, Florida, in *Proc. 4th Annu. Conf. Restor. Coastal Vegetation Fla.*, May 14, 1977, Lewis, R. R. and Cole, D. P., Eds., Hillsborough Community College, Tampa, Fla., 1978, 45.
13. Banner, A., Revegetation and maturation of restored shoreline in the Indian River, Florida, in *Proc. 4th Annu. Conf. Restor. Coastal Vegetation Fla.*, May 14, 1977, Lewis, R. R. and Cole, D. P., Eds., Hillsborough Community College, Tampa, Fla., 1978, 13.
14. Fehring, W. K., Giovenco, C., and Hoffman, W., An analysis of three marsh-creation projects in Tampa Bay resulting from regulatory requirements for mitigation, in *Proc. 6th Annu. Conf. Wetlands Restor. Creation*, May 19, 1979, Cole, D. P., Ed., Hillsborough Community College, Tampa, Fla., 1980, 191.
15. Courser, W. K. and Lewis, R. R., The use of marine revegetation for erosion control on the Palm River, Tampa, Florida, in *Proc. 7th Annu. Conf. Restor. Creation Wetlands*, May 16—17, 1980, Cole, D. P., Ed., Hillsborough Community College, Tampa, Fla., 1981, 125.
16. Hoffman, W. E. and Rodgers, J. A., Cost-benefit aspects of coastal vegetation establishment in Tampa Bay, Florida, *Environ. Conserv.*, 8(1), 39, 1981.
17. Martyn, E. B., A note on the foreshore vegetation in the neighborhood of Georgetown, British Guiana, with special reference to *Spartina brasiliensis*, *J. Ecol.*, 22, 292, 1934.
18. Ranwell, D. S., *Ecology of Salt Marshes and Sand Dunes*, Chapman and Hall, London, 1972.
19. Lewis, R. R., Lewis, C. S., Fehring, W. K., and Rodgers, J. A., Coastal habitat mitigation in Tampa Bay, Florida, Swanson, G., Tech. Coord., Proc. Mitigation Symp., July 16—20, 1979, Tech. Rep. RM-65, U.S. Department of Agriculture, Ft. Collins, Colo., 1979, 136.

Chapter 8

# MANGROVE FORESTS

### Roy R. Lewis, III

## TABLE OF CONTENTS

# I. INTRODUCTION

The protection, restoration, creation, and enhancement of mangrove forests has, in recent years, received a lot of attention.[1-4] The primary reason for this is that the long ignored ecological values of mangrove forests as habitat and detrital food sources were scientifically documented for the first time in a U.S. mangrove forest[5,6] at the same time that ecologists realized that large scale losses of mangroves, due to coastal development, were occurring throughout the world.[7-12] Following the breakthroughs in the early 1970s in techniques for artificially creating tidal marshes (predominantly *Spartina alterniflora* L.)[13] techniques for mangrove forest restoration and creation were developed.[1]

# II. THE NATURAL PLANT COMMUNITY

## A. Distribution

Mangrove forests are intertidal areas along protected tropical and subtropical shorelines which are vegetated with woody plants that may exist as trees, shrubs, or bushes depending on the climate of the region. Chapman[14] recognizes 72 species as being mangroves including one genus of fern and three genera of palms.

Maximum development of the forest system usually occurs in areas of high rainfall, high temperatures, and low incidence of typhoons or hurricanes.[15,16] Forests with 20 + species and maximum canopy heights of 20 to 30 m are found in Southeast Asia, while at the northern or southern limit away from the equator (about 30° of latitude) the occurrence of frost reduces the forest to monospecific stands (usually *Avicennia*) 1 to 2 m in height.

## B. Primary Productivity

Lugo and Snedaker[9] list nine values for net primary productivity from a number of sites in Florida and Puerto Rico. The mean of the nine values (range 0 to 7.50 gC/m²/day) is 3.04 gC/m²/day or 1109.6 gC/m²/year. Thayer et al.[17] lists two values for mangrove primary productivity, 400 gC/m²/year for *Rhizophora mangle* L. and 1022 gC/m²/year for *Avicennia germinans* L. Stearn.

## C. Secondary Productivity

As noted by Lugo and Snedaker[9] "quantitative studies of the secondary productivity of the consumers associated with mangrove ecosystems have not been performed".

Studies have shown, however, that detritus formed from mangrove leaves in South Florida is the basis for a food web that includes many species of invertebrates, fish, birds, and man.[5,6] Similar studies on fish in Puerto Rico[18] have found the same results.

Many authors have found the fauna of mangroves to be abundant and diverse, and often include species important to commercial and sport fisheries throughout the world.[6,15-17]

In addition to providing a source of food, mangroves also serve as habitat for many adult and juvenile fish and invertebrates[5,6] as well as providing nesting sites for many species of colonial waterbirds.[19-21] Woolfenden and Schreiber[22] note that mangrove forests in Florida: "... are absolutely essential to the existence of a large number of water birds that breed in Florida, for essentially all of the breeding colonies of pelicans, cormorants, herons and ibises of saline environs are in mangrove."

MacNae[15] also notes the occurrence of many species of birds in the mangroves of southeast Asia and Chapman[14] lists 48 species of birds and 13 species of mammals recorded from mangrove forests around the world.

## Table 1
## DOCUMENTED MANGROVE LOSSES

| Location | Original areal cover(ha) | Existing areal cover(ha) | Loss (%) | Ref. |
|---|---|---|---|---|
| Florida (U.S.A.) | | | | |
|   Tampa Bay | 10,053 | 5,630 | −44 | 12 |
|   Biscayne Bay | 63,300 | 11,100 | −82 | 10 |
| Puerto Rico | | | | |
|   Main Island | 24,300 | 6,405 | −75 | 35 |
|   Vieques Island | 446 | 367 | −18 | 43 |
| El Salvador | No areal figures given | | −50 | 36 |
| | | | | |
| Australia | | | | |
|   Botany Bay | 1,500 | 1,000 | −33 | 44 |
| South Vietnam | 286,400 | 104,123[a] | −36 | 38 |

[a]  Natural recovery from herbicide damage has not been documented since 1973.

## D. Silvaculture

In addition to their ecological value, mangroves have been used for centuries as sources of firewood, construction timber, salt, tannins, dyes, and even food.[15,23,24]

Because of these many uses, mangroves have been grown and harvested using standard silvaculture practices in the Adaman Islands,[25] Thailand,[26,27] Malaya,[28-30] India,[31] Puerto Rico,[32,33] and Indonesia.[34] Rotation times vary with the species and range from 20 to 100 years. The earliest active attempts to plant mangroves developed from silvaculture practices and will be discussed in later sections.

## E. Historical Losses

Due to overharvesting,[31] dredging and filling for development,[7,10,12,35] various hydrological modifications,[9,36,37] aerial defoliation during the Vietnam war,[38] and various other human impacts including oil spills,[39] tens of thousands of hectares of mangroves have been either permanently destroyed or temporarily impaired. Table 1 lists the available figures for documented losses. The list is very short since most of the data on mangrove losses has not been gathered together in one publication. Such a compilation is presently in progress[90] and will hopefully put the problem in better perspective.

In any case, the losses are assumed to affect fisheries and other faunal components of the food web but little documentation exists. Lewis[12] reports a 20% decline in commercial fisheries catches along Florida's Gulf coast after two peaks of 61,400,000 kg in 1960 and 61,500,000 kg in 1965. During this same period, 40% of the mangroves of one of the main estuaries (Tampa Bay) in the area was lost due to residential and commercial fills. Lindall[40] reports that 85% of the commercial fish and shellfish caught in South Florida is dependent on estuaries like Tampa Bay for at least a portion of their life cycle and that hydrological alterations and habitat loss threatens the $10 million commercial harvest and $575 million sports harvest of estuarine dependent species.

Crowder[41] notes that only 35,000 breeding pairs of native wading birds remain in South Florida from an estimated 2.5 million in 1870. Modifications to and loss of both freshwater and marine wetlands (mangroves) are cited as one of the main causes for this decline.

Daugherty[36] reports a 50% decline in the shrimp catch for El Salvador since 1964 and other declines in reptile, bird, and mammal populations associated with a nearly 50% loss of mangroves for El Salvador.

DeSylva and Michel[42] were unable to demonstrate conclusively that the partial or

### Table 2
### SPECIES OF MANGROVES USED IN PLANTINGS FOR
### SILVACULTURE, MITIGATION, AND EXPERIMENTATION

| Species | Type of Planting | Ref. |
|---|---|---|
| Rhizophoraceae | | |
| *Rhizophora mangle* L. | P, T, AP, N | 1, 2, 4, 23, 33, 45—56 |
| *R. mucronata* Lamk. | P | 23, 30 |
| *R. apiculata* Blume | P | 15, 27, 29, 30, 38, 57 |
| *Bruguiera gymnorrhiza* (L.) Lamk. | P | 23 |
| *B. parviflora* (Robt.) W. and A. ex Griff. | T | 30 |
| *B. sexangula* (Lour.) Poir. | P | 58 |
| Sonneratiaceae | | |
| *Sonneratia caseolaris* (L.) Engler | P | 58 |
| Avicenniaceae | | |
| *Avicennia germinans* (L.) L. | P, T, N | 1, 46, 49—53, 59—61 |
| *A. marina* (Forsk.) Vierh. | T | 62, 63 |
| *A. officinalis* L. | T | 62 |
| Combretaceae | | |
| *Laguncularia racemosa* (L.) Gaertn. f. | T, N | 1, 33, 46, 51—53, 56, 59, 64, 65 |
| Theaceae | | |
| *Pelliciera rhizophora* Plance and Triana | T | 64 |
| Palmae | | |
| *Nypa fruticans* Wurmb | N | 30 |

*Note:* P = Propagules — fresh seeds or seedlings T = Transplants, AP = Aerial drops of propagules, and N = Nurseried seedlings.

complete defoliation of 104,123 ha of mangroves in South Vietnam permanently damaged the estuarine ecology of the area although they noted increased turbidity and erosion due to the lack of vegetative cover and a dramatic decrease in fisheries harvests. It was not possible to separate overfishing impacts from social and defoliation causes.

## III. PLANT MATERIAL TYPES AND SOURCES

Table 2 lists the 13 species of mangroves which have been used in silvaculture, mitigation, or experimentation involving planting mangroves. The 13 species fall into 6 families, each with its own variation on the generally viviparous methods of sexual reproduction shown by mangroves.[14,15]

For the sake of this discussion, the word "propagule" will be used when referring to the seeds or seedlings of a species that are collected directly from the tree, or very soon thereafter, and have not exhibited any additional expansion or root formation. This would correspond to the Type A or B seedling of Teas et al.[47] This is preferable to using "seed" or "seedling" since these words are often used interchangeably for the same thing, depending on the writer's determination as to how viviparous a particular species is. MacNae[15] considers that the members of both the Rhizophoraceae and Avicenniaeceae exhibit "apparent vivipary" while Chapman[14] notes that the Avicenniaceae along with some other groups "differ from the other viviparous mangroves in that the seedling remains enclosed within the testa whilst on the mother plant". Chapman[14] also notes that "retention of the testa inhibits seedling growth in *Avicennia*".

The term "seedling" will be used to apply to propagules that have germinated and show additional changes such as the loss of the testa in *Avicennia* or root growth from the radicle in *Rhizophora*. This would correspond to the types C and D seedlings of Teas et al.[47]

Air-layering has been successful in experiments with *R. mangle, A. germinans,* and *Laguncularia racemosa*[70] but has not been used for any actual production of planting materials. Its ease and cost effectiveness warrant further research. Table 2 indicates the type of plantings that have been tried for each species; propagules (P), transplanting (T), aerial planting (AP), and nursuried seedlings (N). Aerial planting has only been tried with propagules of *Rhizophora*[48]. As can be seen from the table, the use of propagules or transplanting seedlings or small trees have been the two most widely used techniques.

Up until recently, the most available source of plant materials were the forests themselves where propagules can be gathered during certain seasons of the year or seedlings on small trees can be transplanted year-round. More recently, commercial sources of mangrove plant materials have appeared (Appendix 1). These are presently limited to providing the four North American species of mangroves (*R. mangle, A. germinans, L. racemosa,* and *Conocarpus erecta* L.) but other commercial ventures are to be expected.

The seasonal availability of propagules needs to be closely monitored by anyone anticipating their need for mangrove propagules. Several authors have noted the August to October peak in availability of the propagules of the North American species[1,46] and Steinke[71] has noted the main fruiting period for *A. marina* in Natal, South Africa, was March/April. It is obvious that anyone planning to use propagules for a planting should also schedule their installation during the peak availability of propagules, otherwise there will be none available.

## IV. TECHNIQUES USED IN REVEGETATION

As noted before, four techniques have been used: direct planting of propagules, aerial planting of propagules, transplanting of seedlings or small trees, and planting of nursuried seedlings.

Table 3 lists, in approximate chronological order, 34 individual cases where mangrove plantings have occurred. The author is aware of several dozen more in various stages of planning and implementation.

Most of the plantings prior to 1970 were for silvaculture and date back to the 19th century in the Phillipines.[15,27,29,30,33,38,57] These are the largest plantings with individual areas of up to 15,000 ha being reported as being planted.[27] Plantings also were used for erosion control[15,45,58,62] and experimental analysis of mangrove biology.[46,55,64] Beginning with the realization of the ecological value of mangroves and the passage of laws protecting them from destruction, particularly in the U.S., many smaller scale plantings for restoring damaged areas or mitigating environmental damage have occurred.[1-4,47,49-56,59-61,63-69]

Along with the listing of the planting, some comment about the project is included. It is apparent that depending on the technique of planting, the type of plant material used, and the site of the planting, success can vary from 0 to 100%. Several authors have noted that two of the most critical factors in successful projects are (1) a planting site with little or no wave action against the shore to dislodge plantings[1-4,49-55] and (2) proper elevation within the intertidal zone.[1-4,49-55] It is not surprising that these are also critical factors in the success or failure of plantings of tidal marsh plants (see Chapter 5).

Concerning the first factor, wave energy, a number of plantings have been tried along eroding shorelines,[1,54] on shifting sand deposits,[51] or simply on high energy shores to protect them and to see if the plants could survive.[55,60-61,66] Even with some sort of wave barrier[60,61] or erosion protection such as tires,[54] the plantings were nearly 100% unsuccessful. In some of the experiments, simultaneous plantings were also made in low wave energy sites[54,55,66] and much greater success (65 to 90%) occurred.

Table 3

ATTEMPTED PLANTINGS OF MANGROVES IN APPROXIMATE CHRONOLOGICAL ORDER

| Location | Species | Dates | Comment | Ref. |
|---|---|---|---|---|
| India | | | | |
| Adaman Islands | *Rhizophora mucronata* Lamk. | 1898—1908 | Silvaculture, 277 ha | 23 |
| Phillipines | | | | |
| Manila Bay | *R. mucronata* | Mid 19th century to 1910s | Silvaculture | 30 |
| Malaysia | | | | |
| Matang | *R. apiculata* Blume | Since 1900 | Large scale silvaculture of 45,000 ha — 20% of which is planted | 29, 57 |
| Perak and Selangor | *R. mucronata, R. apiculata Nypa fruticans* Wurb. | 1920s | Large scale silvaulture, 100,000 + ha | 30 |
| | *R. sp., Bruguiera parviflora* (Roxb.) W. and A. ex Griff. | | Small scale silvaculture from nurseried plants | 30 |
| | | | Small scale attempted transplants — low survival | 30 |
| Hawaii | *R. mangle* (L) | 1902 | Introduced seedlings planted on mudflats to control erosion | 58 |
| | *R. mucronata, B. sexanula* (Lour.) Poir., *Sonneratia caseolaris* (L.) Engler | 1922 | Introduced from the Phillipines | 58 |
| U.S. | | | | |
| Florida Keys | *R. mangle* | 1915—16 | Planted to control erosion adjacent to the overseas highway | 45 |
| Jewfish Creek (Monroe County) | *R. mangle, Avicennia germinans* (L.) L., *Laguncularia racemosa* (L) Gaertn. f. | 1938 | 200 Experimental transplants — only 7 survived after 9 months | 46 |
| Dry Tortugas | *R. mangle* | 1938 | 4100 fresh propagules — 3300 survived to July 1939; none found in 1970 | 1, 46 |
| South Vietnam | | | | |
| Ca-Mau Peninsula | *R. apiculata* | 1930s to date | Long scale silvaculture 38,000 ha | 38 |
| Puerto Rico | *R. mangle, A. germinans, L. racemosa* | 1930s | *Rhizophora* planted from seed; *Avicennia/Laguncularia* transplanted | 33 |
| Thailand | *R. apiculata R. mucronata* | 1946 | Large scale silvaculture 15,300 ha planted | 27 |

| Location | Species | Date | Notes | Ref. |
|---|---|---|---|---|
| Java | *A. marina* (Forsk.) Vierh., *A. officinalis* L. | 1950s | Planted to stabilize fish ponds | 62 |
| Ceylon | *R. apiculata* | 1960s | Planted to encourage deposition of silt | 15 |
| Bahamas, Exuma | *R. mangle* | 1968—69 | Small scale experiments 0—71% survived depending on wave regime | 55 |
| Panama | *Rhizophora* sp., *Pelliciera rhizophorae* Planch and Triana, *L. racemosa*, *Avicennia* sp. | 1960s | Small scale experimental transplants of seedlings — good survival | 64 |
| U.S. Tampa Bay and Sarasota Bay, Fla. | *R. mangle*, *A. germinans*, *L. racemosa* | 1969—71 | Large scale experimental plantings and transplantings | 52—53 |
| Tampa Bay, Fla. | *R. mangle*, *A. germinans*, *L. racemosa* | 1973 | 40 of each species (0.5—1.5 m high) were transplanted successfully | 56 |
| Marco Island, Fla. | *R. mangle*, *A. germinans*, *L. racemosa* | 1973 | 2447 transplants (15.7% survival) on dredged material | 51 |
| Texas | *A. germinans* | 1974—75 | Transplants of 0.10—0.75 in high trees 0—17% survival | 60, 61 |
| St. Lucie Inlet, Fla. | *R. mangle* | 1975 | Transplants of 4—6 year old trees 65—85% survival | 54 |
| Charlotte County, Fla. | *R. mangle* | 1975 | 2.23 ha planted with 60,000 propagules 85—90% survival. | 47 |
| St. Lucie County, Fla. | *R. mangle* | 1975 | 3 sites — 2628 propagules and seedlings planted; 0—90% survival | 47 |
| Siesta Key, Fla. | *L. racemosa* | 1976 | Transplants 0.75—1.0 m high — 100% survival | 65 |
| Miami | *L. racemosa*, *A. germinans* | Mid-1970s | Transplants of 14 trees to 6 m high — no survival after 6 months | 1 |
| Tampa Bay, Fla. | *R. mangle* | 1975 | Small scale experimental plantings of propagules in 3 wave energy zones — only protected plantings survived. | 66 |
| Key West, Fla. | *R. mangle* | 1977 | Propagules and seedlings showed about 45% survival — 2—3 year old trees transplanted showed 98% survival | 3 |
| Miami | *R. mangle* | 1977 | Experimental aerial planting | 48 |
| St. Croix, U.S. Virgin Islands | *R. mangle*, *A. germinans* | 1978—79 | 86,000 *R. mangle* propagules and 36,000 *A. germinans* propagules planted in oil damaged site (6.15 ha); 40% survival of *Rhizophora* and 1—2% survival of *Avicennia* | 49, 50 |

## Table 3 (continued)
## ATTEMPTED PLANTINGS OF MANGROVES IN APPROXIMATE CHRONOLOGICAL ORDER

| Location | Species | Year | Description | Ref. |
|---|---|---|---|---|
| **U.S.** | | | | |
| Tampa Bay, Fla. | A. germinans, L. racemosa | 1979 | 1513 transplants (0.3—1.9 m tall) 73.3 survival after 13 months | 59, 67 |
| Punta Gorda, Fla. | R. mangle | 1980 | 2.83 ha planted with 35,000 propagules 70% survival | 68 |
| **Australia** | | | | |
| Sydney, New South Wales | A. marina | 1980—81 | Transplanting of 0.5—1.0 m trees; essentially 100% survival in low wave energy, 0% in medium wave energy | 63 |
| **U.S.** | | | | |
| Key Largo, Florida | R. mangle, A. germinans | 1981 | 2 ha of R. mangle propagules 90 + % survival — 0.75 ha of A. germinans propagules — 25 + % survival 2.25 ha of nursuried A. germinans seedlings — (1% survival) | 69 |

The only reported success at establishing mangroves on an eroding shoreline is that of Goforth and Thomas.[3] At their highest wave energy site, propagules and 12- to 18-month-old seedlings of the red mangrove showed very poor survival. Some 2- to 3-year-old small trees (0.4 to 0.8 m tall), however, showed excellent (98%) survival after 23 months. The use of a power auger to provide a hole for planting was, no doubt, a deciding factor for this high success.

Concerning proper elevation of plantings, it is important to first determine the general intertidal zone elevations which will depend upon the tidal range. Once this general zone is delineated, the best zone for each species is then determined. The easiest way to do this is simply survey the elevations of existing mangroves at the closest location to the proposed planting site. In general, the Rhizophoraceae can be planted in zones of greater inundation than species found in the upper intertidal (*Avicennia, Laguncularia*). Rabinowitz,[64] utilizing transplant experiments, tested the survival of four species at elevations different from that where they are normally found. *L. racemosa* did not do well in the lower elevations normally occupied by *Rhizophora* spp. and *Peliciera rhizophorae*, but *Avicennia germinans* did appear to do well in all zones. Goforth and Thomas[3] found that survival of both propagules and seedlings of *R. mangle* was twice as great at the +0.1 m tidal level as at a 0.0 m elevation. Teas et al.[47] noted that "elevation with respect to tidal levels was a significant factor in mangrove establishment" and that *A. germinans* and *L. racemosa* did not naturally establish among planted *R. mangle* below +0.4 m mean sea level but above that elevation gradually became more common as "volunteers" or naturally established (no planting) seedlings from water borne propagules. The importance of understanding this natural secondary succession is discussed in a later section.

Regarding transplanting Pulver[56] states that "each tree regardless of species should be planted at an elevation similar to that at which it originally grew."

Regarding what type of plant material propagule, seedling, transplant, or nursuried tree) to be used, a balance between cost, expected success, and time lapse until the planting is mature must be struck. Table 4 lists the expected costs of various methods of planting mangroves. The amounts vary from $1140 to $216,130/ha depending on the plant material used and the spacing of the installations. It is apparent that for a given spacing, the costs increase substantially from the lower end, using propagules, to the higher end, using larger trees. It is also apparent that spacing is also a critical factor. It is important for anyone recommending a plant spacing to understand that reducing the distance by ⅓ (0.91 m [3 ft] to 0.61 m [2 ft]) more than doubles (12,100 to 26,896) the number of installations required and a further reduction to 0.30 m (1 ft) spacing increases it to 110,889. Failure to understand this has caused many costly misunderstandings. The increased costs of growing propagules to seedlings or trees is the other cost factor.

If it is determined that a damaged or cleared forest needs to be put back in its original form immediately, then larger trees can be moved. Teas[1] reports no success in moving *A. germinans* and *L. racemosa* trees up to 6 m tall while Pulver[56] indicates that five of six *R. mangle* (4.5 to 6.5 m tall) moved by Gill[71] in 1971 were thriving in 1973. Two factors rule this method out for general use. The first is the cost (as yet undetermined — but probably high) and the second is the availability of donor sites. The only time this might be useful would be to "salvage" larger trees that are to be destroyed by development. The general rules for transplanting mangroves as outlined by Pulver[56] should be followed. They are summarized below for 0.5 to 1.5 m pruned mangroves.

1. Top and side branches should be pruned to ⅔ their original length.
2. Trees should be removed with a root ball diameter about half the original tree height.

## Table 4
## ESTIMATED COST ($/ha) FOR PLANTING MANGROVES BY USING VARIOUS TECHNIQUES

| Species and technique | Spacing (m) | | | Ref. |
|---|---|---|---|---|
| | 0.30 | 0.61 | 0.91 | |
| *Rhizophora Mangle* | | | | |
| Propagules (collected) | 10,175 | 2,470 | 1.140 | 1 |
| | | | 12,500[a] | 49 |
| | | | 6,250 | 49 |
| | 26,000 | 13,000 | 6,545[a] | 4 |
| Propagules (purchased) | 11,251 | 2,742 | 1,261 | 1 |
| | 30,000 | 14,000 | 7,000 | 4 |
| *R. mangle, Avicennia germinans, Laguncularia racemosa* | | | | |
| 6-month-old seedlings | 22,400 | 5,400 | 2,510 | 1 |
| (purchased) | 107,593 | 27, 2 32 | 12,103 | 4 |
| 3-year-old trees (purchased) | | | 216,130 | 1 |
| | | | 40,755 | 4 |
| | | | 70,000 | 4 |
| *R. mangle* | | | | |
| 3-year-old trees (transplanted) | | | 45,386 | 3 |
| *A. germinans, L. racemosa* | | | | |
| (transplanted) | | | 11,459 | 59 |

[a]   Actual cost of commercial project.

Modified from Lewis, R. R., Proc. U.S. Fish Wildl. Serv. Workshop Coastal Ecosystems Southeastern U.S., Markouts, P. S., Ed., February 18—22, 1980, Big Pine Key, Fla., U.S. Fish and Wildlife Service, Washington, D.C., 1981, 88.

3.   The rootball should be watered and stamped down while replacing soil to and sealing between the rootball and the sides of the hole.

4.   Trees should be replanted at approximately the same level in the ground and at approximately the same tidal elevation as in the original habitat.

5.   Trees should never be planted in unstable substrates.

Previous experience at transplanting from nature of 0.5 to 2.0 m trees has met with varying success. Evans[65] had 100% success with *L. racemosa*; Pulver[56] had essentially the same results with *R. mangle, A. germinans,* and *L. racemosa,* while Hoffman and Rodgers[59] reported a 73.3% survival for *L. racemosa* and *A. germinans.* All these transplants were less than 2 m in height. Kinch[51] on the other hand had only 15.7% survival with these same species on unstable dredged material and Watson[30] indicates transplanting of *Rhizophora* spp. and *Bruguirera parviflora* in Malaysia "have been only partially successful... in spite of great care in planting." Gibbs[68] has had variable success transplanting *A. marina* with no survival along eroding or higher wave energy shores but essentially 100% survival in protected areas.

Under most circumstances to date, the decision on what type of plant material to use has led to the use of propagules or seedlings because of their ready availability, low cost, and ease of installation. Propagules, in particular, have been used widely and over 50 years ago, the following comments about use of propagules of *Rhizophora* spp. and *Bruguirera* spp. were written: "Where seed-bearers of the Rhizophoras are wanting, it may be necessary to collect germinating seeds {propagules} from trees nearby, and stick them in the mud... care should be taken to select healthy seed {propagules} for the purpose, and to use only those that have recently fallen from the trees, or that will come away without pulling... the seeds should be struck into a depth of a few inches only, so that they will not fall over; deep insertion is not recommended... the seedlings {propagules} are thrust into the mud at intervals from 40-100 centimeters... young plantations are protected from damage by floating objects."[30]

These recommendations are equally valid today when used with the previous comments about tidal elevations of plantings and recognition of the wave energy regime of the site.

Figure 1 shows a mangrove restoration site in St. Croix, U.S. Virgin Islands, where 86,000 *R. mangle* propagules were installed in August 1978. Figure 2 is the same area in April 1980. The site was a mangrove forest that had been damaged by an oil spill in 1971.[39,49-50] Survival after 20 months and 2 hurricanes was 40%. The main causes of loss of propagules or seedlings after the propagules germinated were[49]

1. Physical removal due to erosion, accumulations of seagrass wrack, or floating debris.
2. Eating of the planted seeds by unknown biological agents, possibly crabs.
3. Death of seedlings due to natural causes or residual oil.
4. Apparent planting at too high an elevation.

Figure 3 shows a typical seedling that has been attacked by some organism. Watson[30] also reports on problems with crabs eating the propagules or monkeys pulling them out and Savage[52] notes that marsh rabbits have been seen eating *R. mangle* propagules.

In addition to hand-planting of propagules, aerial planting by dropping the propagules enclosed within a bag with stabilizing streamers, called a "missile", has been tried experimentally in Florida and Vietnam with some success.[48] The main question about the feasibility of using this method involves the cost differential between hand planting and use of "missiles". As noted by Teas and Jurgens,[48] the method has potential for use for establishing mangroves at isolated sites not easily reached on foot or by vehicle and could be used with other members of the Rhizaphoraceae including *Ceriops, Bruquiera*, and *Kandelia*.

Direct planting of propagules of other species has not met with much success. Lewis and Haines[50] report that only 1 to 2% of their 32,000 propagules of *A. germinans* became established due to the tendency for them to be carried by the tides away from the site where they were broadcast. Those that did become established, however, grew quite well (Figure 4). Steinke[71] has noted that propagules of *A. marina* will not germinate unless they have enough moisture to shed the seedcoat (testa). The same property has been seen in *A. germinans*.[72] For this reason, if the *Avicennia* propagule is placed in the ground with the seed coat on, its chances of success are much less than if the testa is allowed to naturally drop off or is manually removed.

For species other than those in the Rhizophoraceae, it appears that germination of the propagule in a nursery to produce a potted plant that is then planted after a short growing period (usually 3 to 6 months) is presently the most viable option for using those species. This technique, using either nursuried seedlings or naturally established ones that were transplanted, has been used with *R. mangle*.[3,47] The evidence to date indicates that "rooted seedlings had no advantage over unrooted propagules"[47] and "seedlings showed no advantage over propagules in terms of growth or transplant".[3] In addition, as can be seen in Table 4, the cost of using seedlings of *Rhizophora* is 2 to 3 times that of using propagules.

Nursuried seedlings of the other species have not had widespread use due to their limited availability but in small scale experimental use they work quite well.[73] The

FIGURE 1.   Mangrove planting area on the island of St. Croix, U.S. Virgin Islands, August 1978. Red mangrove (*Rhizophora mangle* L.) propagules have been inserted in the ground on 0.8 to 1.0 centers.

FIGURE 2.   Mangrove planting area on the island of St. Croix, U.S. Virgin Islands, April 1980. Same area as Figure 1. Mangrove propagules have now germinated and are growing.

FIGURE 3. Unsuccessful red mangrove (*Rhizophora mangle* L.) seedlings that have had their upper portions grazed by some unknown animal.

FIGURE 4. Black mangrove (*Avicennia germinans* (L.) L.) approximately 1.0 m tall that has grown from broadcast propagules; age 20 months. St. Croix, U.S. Virgin Islands, April 1980.

commercial suppliers in Appendix 1 generally have seedlings of *A. germinans* and *L. racemosa* available year round. Anyone anticipating growing seedlings of mangroves is encouraged to thoroughly familiarize themselves with the extensive literature on experimental culture of mangroves[71,74-80] to determine the best methods to be used.

## A. Mangrove Forest Succession

The classic division of succession into primary and secondary types if attributed to Clements:[81] "... all bare areas fall into... primary and secondary areas. Primary areas present extreme conditions... possess no viable germules of other than pioneer species... and hence give rise to long and complex seres. Secondary bare areas present less extreme conditions, normally possess viable germules of more than one stage, often in large numbers... and give rise to relatively short and simple seres." Chapman[82] describes nine patterns of primary succession for mangroves and notes one secondary succession pattern beause of silvacultural practices in India.

It is important to understand these patterns of succession for a given area because natural patterns of colonization of bare areas such as a new dredged material island (primary succession) or regrowth after clear-cutting (secondary succession) are generally not preventable, except by extraordinary means, and can either assist in the revegetation process or inhibit it. Heavy emphasis on planting an area with a species that will naturally flood the site with propagules and revegetate quickly and naturally is obviously a waste of time and money. Knowing which species will *not* return quickly by themselves gives direction to any wetland restoration or creation effort.

Actual observation of primary succession in mangroves has generally been limited to shallow or emerged sand bars,[46] man-made dredged material deposits,[83,84] or seagrass meadows.[46,77]

Shallow grass beds or sand bars are usually colonized by *Rhizophora* or a related species with a long radicle (up to 1 m). Man-made dredged material deposits are, conversely, usually colonized by upper intertidal species such as *L. racemosa*[83,84] or grasses such as *Spartina alterniflora*[46,83] or *S. brasiliensis.*[82]

Secondary succession in mangrove forests has been studied by Holdridge[33] and Wadsworth[32] in Puerto Rico; Durant,[28] Watson,[30] and Noakes[29] in Malaya; Banijbatana[27] in Thailand; Detweiler et al.[5] in Florida; and MacNae[15] in the Indo-Western Pacific. In nearly all situations, the species that first colonize and dominate a recently cleared forest are not the dominant species of the mature forest. This is the classic case of several seres leading to a climax community. Detweiler et al.[5] compared an undisturbed mangrove forest with an adjacent forest that had been cleared in Tampa Bay, Fla. They found that *Salicornia virginica* L. and *Spartina alterniflora* Loisel were the dominant plants in the sere present 3 years after disturbance. *Rhizophora*, in particular, were not recolonizing the site. As a result, the developer who originally cleared the site was required to plant approximately 25,000 *R. mangle* propagules. Part of the reason this species may not have been recolonizing was the presence of dead trees and slash from the original clearing. MacNae[15] mentions this same problem.

MacNae[15] also mentions the ''nurse'' effect of one species on another, specifically involving *A. marina* in Natal and *B. parviflora* in Australia and Burma. The nurse effect is described as: ''Once established the *Avicennias* and the species of *Sonneratia* cause accretion by impeding water movement, the soil level rises, and other species of *Rhizophora, Bruguiera,* and *Xylocarpus,* germinate from stranded seeds and become established. All these trees tend to grow taller than the pioneers, over-top them and these then die off. Hence one rarely finds a well-grown tree of *Avicennia marina* in a *Rhizophora* or *Bruguiera* forest.''

Watson[30] also notes that the marsh fern (*Achrostichum aureum* L.) can act as a nurse in its small form or as a competitor in its large form.

Facilitation of establishment of a species by another species has not received as much study as exclusion phenomena. An example is the invasion of broom-sedge (*Andropogon* spp.) into old fields facilitated by shading of its seeds by tall weed flora (*Solidago* spp., *Aster* spp.).[86] Lewis and Dunstan[87] have described such facilitation by *Spartina alterniflora,* which creates a physical trap to hold seedlings of red, black, and white mangroves. Pioneer *S. alterniflora* marshes on dredged material islands in Tampa Bay are thus gradually replaced by mangrove forests. These forests are dominated by black and white mangroves, possibly as a result of selective exclusion of larger red mangrove seedlings.[88]

Exclusion of one species by the establishment of another is widely noted. For example, Niering and Egler[89] describe the exclusion of trees by a community dominated by a shrub (*Viburnum lentago*). Similarly, Holdridge[33] mentions the physical exclusion of mangrove seeds in marsh fern (*A. aureum*) areas in cutover mangrove areas in Puerto Rico. MacNae[15] mentions similar exclusion by slash, *Acrostichum* spp. or

## Table 5
## MANPOWER ESTIMATES FOR MANGROVE PLANT MATERIAL COLLECTION AND INSTALLATION

| Species | Plant material | Task | Spacing | Man-hours | Ref. |
|---|---|---|---|---|---|
| *Rhizophora mangle* | Propagules | Collection and installation | 0.8—1.0 m | 1828/ha | 49 |
| | Propagules | Collection only | — | 400—1000/hr | 48 |
| | 0.4—0.8 m tall trees | Transplanta-tion | 1.0 m | 3098/ha | 3 |
| *R. mangle, Avicennia germinans* | Propagules | Collection and installation | 0.8—1.0 m | 457/ha | 50 |
| *A. germinans, Laguncularia racemosa* | 0.3—1.9 m tall trees | Transplanta-tion | 1.0 m | 2541/ha | 59 |
| *Spartina alterniflora* Loisel. | 0.5—1.0 m | Transplanta-tion | 1.0 m | 995/ha | 59 |

beach thistle (*Acanthus*). Lewis[49] attributed the very slow recovery of an oil damaged mangrove forest on St. Croix, at least partially, to physical exclusion of red mangrove seedlings by dead prop roots and fallen limbs. Watson[30] and MacNae[15] have both noted the same phenomenon.

The importance of understanding the natural succession in mangrove forests and the nurse and exclusion phenomena is that each forest system has its own unique characteristics that may indicate: (1) natural recovery will be sufficient to provide revegetation and that manual planting is unnecessary, (2) a "nurse" species may be the best species to use in revegetation, or (3) problems of exclusion by other plant species or slash may require large-scale clearing of the site before planting.

### B. Manpower Estimates
Very limited information is available on manpower estimates for various phases of collection and installation of mangrove plant materials.[3,48-50,59] These are summarized in Table 5.

From this limited data, it can be seen that collection and installation of propagules (457 to 1828 man-hours/ha) requires about one half the time that transplantation of small (<2 m) trees (2541 to 3098 man-hours/ha). For comparison, the manpower requirements to use smooth cordgrass (*S. alterniflora*) as a substitute for mangroves are 995 man-hours/ha. As mentioned before, this species is a nurse plant for mangroves in Florida.[87,88]

### C. Recommendations for Future Research
Future research in mangrove forest restoration and creation should concentrate on these areas:

1. Reduction of planting costs through innovative techniques such as aerial planting[48]
2. Greater experimentation with seedlings of species other than those in the Rhizophoraceae
3. Application of horticultural selection techniques to develop lines of species adapted for various latitudes
4. More quantitative analysis of ongoing planting projects including use of control plots to determine rates of natural seedling recruitment

# REFERENCES

1. **Teas, H. J.**, Ecology and restoration of mangrove shorelines in Florida, *Environ. Conserv.*, 4, 51, 1977.
2. **Teas, H. J.**, Restoration of mangrove systems. Proc. U.S. Fish Wildl. Serv. Workshop Coastal Ecosystems Southeastern U.S., Markouts, P. S., Ed., February 18—33, 1980, Big Pine Key, Fla., U.S. Fish and Wildlife Service, Washington, D.C., 1981, 95.
3. **Goforth, H. W. and Thomas, J. R.**, Plantings of red mangroves (*Rhizophora mangle* L.) for stabilization of marl shorelines in the Florida Keys, in *Proc. 6th Annu. Conf. Wetlands Restor. Creation,* Cole, D. P., Ed., May 19, 1979, Hillsborough Community College, Tampa, Fla., 1980, 207.
4. **Lewis, R. R.**, Economics and feasibility of mangrove restoration, in Proc. U.S. Fish Wildl. Serv. Workshop Coastal Ecosystems Southeastern U.S., Markouts, P. S., Ed., February 18—22, 1980, Big Pine Key, Fla., U.S. Fish and Wildlife Service, Washington, D.C., 1981, 88.
5. **Odum, W. E.**, Pathways of energy flow in a south Florida estuary, Sea Grant Tech. Bull. No. 7, University of Miami, Coral Gables, Fla., 1979.
6. **Odum, W. E. and Heald, E. J.**, Trophic analyses of an estuarine mangrove community, *Bull. Mar. Sci.*, 22(3), 671, 1972.
7. **Linden, O. and Jernelov, A.**, The mangrove swamp — an ecosystem in danger, *Ambio,* 9, 81, 1980.
8. **UNESCO**, The mangrove ecosystem: human uses and management implications, UNESCO reports in Marine science, 8, UNESCO, Paris, 1980.
9. **Lugo, A. E. and Snedaker, S. C.**, The ecology of mangroves, *Annu. Rev. Ecol. Syst.*, 5, 39, 1974.
10. **Harlem, P. W.**, Aerial photographic interpretation of the historical changes in Northern Biscayne Bay, Florida: 1925 to 1976, Sea Grant Tech. Bull. # 40, University of Miami, Coral Gables, Fla., 1979.
11. **Lugo, A. E. and Cintron, G.**, The mangrove forests of Puerto Rico and their management, in *Proc. Int. Symp. Biol. Manage. Mangroves,* Walsh, G., Snedaker, S., and Teas, H., Eds., October 8—11, 1974, Food and Agricultural Sciences, University of Florida, Gainesville, 1975, 825.
12. **Lewis, R. R.**, Impacts of dredging in the Tampa Bay estuary, 1876—1976, in *Proc. 2nd Annu. Conf. Coastal Soc.: Time-stressed Coastal Environ. Assessment and Future Action,* Pruit, E., Ed., Coastal Society, Arlington, Va., 1977, 31.
13. **Woodhouse, W. W., Jr., Seneca, E. D., and Broome, S. W.**, Propagation of *Spartina alterniflora* for substrate stabilization and salt-marsh development, TM-46, U.S. Army Coastal Eng. Res. Cent., Fort Belvoir, Va., 1974.
14. **Chapman, V. J.**, *Mangrove Vegetation,* J. Cramer, Vaduz, Germany, 1976.
15. **MacNae, W.**, A general account of the fauna and flora of mangrove swamps and forests in the Indo-Western Pacific region, *Adv. Mar. Biol.,* 6, 74, 1968.
16. **Pool, D. J., Snedaker, S. C., and Lugo, A. F.**, Structure of mangrove forests in Florida, Puerto Rico, Mexico and Costa Rica, *Biotropica,* 9(3), 195, 1977.
17. **Thayer, G. W., Stuart, H. H., Kenworth, W. J., Ustach, J. F., and Hall, A. B.**, Habitat values of salt marshes, mangroves, and sea grasses for aquatic organisms, in *Wetland Functions and Values: the State of Our Understanding,* American Water Resources Assoc., Minneapolis, Minn. 1978, 235.
18. **Austin, H. and Austin, S.**, The feeding habitats of some juvenile marine fishes from the mangroves in western Puerto Rico, *Caribb. J. Sci.,* 11(3/4), 171, 1971.
19. **Cawkell, E. M.**, The utilization of mangroves by African birds, *Ibis,* 106, 251, 1964.
20. **Haverschmidt, F.**, The utilization of mangroves by South American birds, *Ibis,* 197, 540, 1965.
21. **French, R. P.**, The utilization of mangroves by birds in Trinidad, *Ibis,* 108, 423, 1966.
22. **Woolfenden, G. E. and Schreiber, R. W.**, The common birds of the saline habitats of the eastern Gulf of Mexico: their distribution, seasonal status, and feeding ecology, *a Summary of Knowledge of the Eastern Gulf of Mexico,* Jones, J., Ring, R. E., Rinkel, M. O., and Smith, R. E., Eds., State University System of Florida, Institute of Oceanography, St. Petersburg, 1-22 of III J, 1973.
23. **Walsh, G. E.**, Mangroves: a review, in *Ecology of Halophytes,* Reimold, R. J. and Queen, W. H., Eds., Academic Press, New York, 1974, 51.
24. **Teas, H. J.**, *Silvaculture with Saline Water,* Vol. 14, Environ. Sci. Ser., Plenum Press, New York, 1979, 117.
25. **Banerji, J.**, The mangrove forests of the Andamans, *Trop. Silvic.,* 20, 319, 1958.
26. **Walker, F. S.**, Regeneration of klang mangroves, *Malay. For.,* 7, 71, 1938.
27. **Banijbatana, D.**, *Mangrove forests in Thailand,* Proc. 9th Pacific Sci. Congr., Bangkok, Vol. II, Forest Resources, 1958, 22.
28. **Durant, C. C. L.**, The growth of mangrove species in Malaya, *Malay. For.,* 10, 3, 1941.
29. **Noakes, D. S. P.**, Methods of increasing growth and obtaining natural regeneration of the mangroves type in Malaya, *Malay. For.,* 18, 22, 1955.
30. **Watson, J. G.**, Mangrove forests of the Malay Peninsula, *Malayan For.,* Rec. No. 6, Fraser and Neave, Singapore, 1928.

31. **Blasco, F.,** *The Mangroves of India,* All India Press, Sri Aurobindo Ashram, Pondicherry, India, 1975.
32. **Wadsworth, F. H.,** Growth and regeneration of white mangroves in Puerto Rico, *Caribb. For.,* 20, 38, 1959.
33. **Holdridge, L. R.,** Some notes on the mangrove swamps of Puerto Rico, *Caribb. For.,* 1(4), 19, 1940.
34. **Versteegh, F.,** *Problems of Silviculture and Management of Mangrove Forests in Indonesia,* FAO Asia — Pacific For. Comm. 77, 1952.
35. **Martinez, R., Cintron, G., and Encarnacion, L. A.,** Mangroves in Puerto Rico: a structural inventory, U.S. Office of Coastal Zone Management, Washington, D.C., 1979.
36. **Daugherty, H. E.,** Human impact on the mangrove forests of El Salvador, in *Proc. Int. Symp. Biol. Manage. Mangroves,* Walsh, G., Snedaker, S., and Teas, H., Eds., October 8—11, 1974, Honolulu, Hawaii, Institute of Food and Agricultural Sciences, Gainesville, Fla., 1975, 816.
37. **Pannier, F.,** Mangroves impacted by human-induced disturbances: a case study of the Orinoco delta mangrove ecosystem, *Environ. Manage.,* 3(3), 205, 1979.
38. **Ross, P.,** The mangrove of South Vietnam: the impact of military use of herbicides, in *Proc. Int. Symp. Biol. Manage. Mangroves,* Walsh, G., Snedaker, S., and Teas, H., Eds., October 8—11, 1974, Honolulu, Hawaii, Institute of Food and Agricultural Sciences, Gainesville, Fla., 1975, 695.
39. **Lewis, R. R.,** Impact of oil spills on mangroves, *Proc. 2nd Int. Symp. Biol. Manage. Mangroves,* July 20 to August 2, 1980, Papua, New Guinea, in press.
40. **Lindall, W. N.,** Alterations of estuaries of South Florida: a threat to its fish resources, *Mar. Fish. Rev.,* 35(10), 26, 1973.
41. **Crowder, J. P.,** Some perspectives on the status of Aquatic wading birds in South Florida, Ecol. Rep. No. DI-SFEP-74-29, U.S. Department of the Interior, Atlanta, Ga., 1974.
42. **deSylva, D. P. and Michel, H. B.,** Effects of defoliation on the estuarine ecology of South Vietnam, in *Proc. Int. Symp. Biol. Manage. Mangroves,* Walsh, G., Snedaker, S., and Teas, H., Eds., October 8—11, 1974, Honolulu, Hawaii, Institute of Food and Agricultural Sciences, Gainesville, Fla., 1975, 710.
43. **Lewis, R. R., Lombardo, R., Sorrie, B., Delusio-Guierri, G., and Callahan, R.,** The mangrove forests of Vieques, Puerto Rico, A report to the U.S. Navy, 1981.
44. **State Pollution Control Commission (New South Wales),** Environmental control study of Botany Bay, Summary rep., 1980.
45. **Bowman, H. H. M.,** Ecology and physiology of the red mangrove, *Proc. Am. Phil. Soc.,* 61, 589, 1917.
46. **Davis, J. H., Jr.,** The ecology geologic role of mangroves in Florida, Papers from Torugas Lab 32, *Carnegie Inst. Washington Publ.,* 517, 305, 1940.
47. **Teas, H. J., Jurgens, W., and Kimball, M. C.,** Plantings of red mangroves (*Rhizophora mangle* L.) in Charlotte and St. Lucie counties, Florida, in *Proc. 2nd Annu. Conf. Restor. Coastal Vegetation Fla.,* Lewis, R. R., Ed., May 17, 1975, Hillsborough Community College, Tampa, Fla., 1976, 132.
48. **Teas, H. J. and Jurgens, W.,** Aerial planting of *Rhizophora mangle* propagules in Florida, in *Proc. 5th Annu. Conf. Restor. Coastal Vegetation Fla.,* Lewis, R. R. and Cole, D. P., Eds., May 13, 1978, Hillsborough Community College, Tampa, Fla., 1979, 1.
49. **Lewis, R. R.,** Large scale mangrove restoration on St. Croix, U.S. Virgin Islands, in *Proc. 6th Annu. Conf. Wetlands Restor. Creation,* Dole D. P., Ed., May 19, 1979, Hillsborough Community College, Tampa, Fla., 1980, 231.
50. **Lewis, R. R. and Haines, K. C.,** Large scale mangrove restoration on St. Croix, U.S. Virgin Islands - II, in *Proc. 7th Annu. Conf. Restor. Creation Wetlands,* Cole, D. P., Ed., May 16—17, 1980, Hillsborough Community College, Tampa, Fla., 1981, 137.
51. **Kinch, J. C.,** Efforts in marine revegetation in artificial habitats, in *Proc. 2nd Annu. Conf. Creation Wetlands,* Lewis, R. R., Ed., May 16—17, 1980, Hillsborough Community College, Tampa, Fla., 1981, 102.
51a. **Kinch, J. C.,** Efforts in marine revegetation in artificial habitats, in *Proc. 2nd Annu. Conf. Restor. Coastal Vegetation Fla.,* Lewis, R. R., Ed., May 17, 1975, Hillsborough Community College, Tampa, Fla., 1976, 102.
52. **Savage, T.,** Florida mangroves as shoreline stabilizers, Florida Department of Natural Resources, Prof. Papers Ser. No. 19, 1972.
53. **Savage, T.,** The 1972 experimental mangrove planting — an update with comments on continued research needs, in *Proc. 5th Annu. Conf. Restor. Coastal Vegetation Fla.,* May 13, 1978, Hillsborough Community College, Tampa, Fla., 1979, 43.
54. **Hannan, J.,** Aspects of red mangrove reforestation in Florida, in *Proc. 2nd Annu. Conf. Restor. Coastal Vegetation Fla.,* Lewis, R. R., Ed., May 17, 1975, Hillsborough Community College, Tampa, Fla., 1976.

55. **Wilcox, L. V., Yocom, T. G., Goodrich, R. C., and Forbes, A.,** Ecology of mangroves in the Jewfish chain, Exuma, Bahamas, in *Proc. Int. Symp. Biol. Manage. Mangroves,* Walsh, G., Snedaker, S., and Teas, H., Eds., October 8—11, 1974, Honolulu, Hawaii, Institute of Food and Agricultural Sciences, Gainesville, Fla., 1975, 305.
56. **Pulver, T. R.,** Transplant techniques for sapling mangrove trees, *Rhizophora mangle, Laguncularia racemosa,* and *Avicennia germinans,* in Florida, Florida Department of Natural Resources, Fla. Mar. Res. Publ., No. 22, 1976.
57. **Diemont, W. H. and Van Wijngaarden, W.,** Sedimentation patterns, soils, mangrove vegetation, and land use in tidal areas of West-Malaysia, in *Proc. Int. Symp. Biol. Manage. Mangroves,* Walsh, G., Snedaker, S., and Teas, H., Eds., October 8—11, 1974, Honolulu, Hawaii, Institute of Food and Agricultural Scienes, Gainesville, Fla., 1975, 513.
58. **Walsh, G. E.,** An ecological study of a Hawaiian mangrove swamp, Lauff, G. H., Ed., Estuaries. Publ. No. 83, American Association for the Advancement of Science, Washington, D.C. 1976, 420.
59. **Hoffman, W. E. and Rodgers, J. A.,** Cost-benefit aspects of coastal vegetation establishment in Tampa Bay, Florida, *Environ. Conserv.,* 8(1), 39, 1981.
60. **Webb, J. W. and Dodd, J. D.,** Shoreline plant establishment and use of a wave-stilling device, Misc. Rept. No. 78-1, U.S. Army Corps Engineering, Coastal Eng. Res. Cent., Fort Belvoir, Va., 1978.
61. **Dodd, J. D. and Webb, J. W.,** Establishment of vegetation for shoreline stabilization in Galveston Bay, Misc. Paper No. 6-75, U.S. Army Corps Engineering, Coastal Eng. Res. Cent., Fort Belvoir, Va., 1975.
62. **Schuster, W. H.,** *Fish Culture in Brackish Water Ponds of Java,* Spec. Pub. Indo-Pacific Fish. Coun. No. 1, 1952.
63. **Gibbs, P.,** New South Wales State Fisheries, Sydney, Australia, personal communication, 1981.
64. **Rabinowitz, D.,** Planting experiments in mangrove swamps of Panama, *Proc. Int. Symp. Biol. Manage. Mangroves,* Walsh, G., Snedaker, S., and Teas, H., Eds., October 8—11, 1974, Honolulu, Hawaii, Institute of Food and Agricultural Sciences, Gainesville, Fla., 1975, 385.
65. **Evans, R. K.,** Techniques and seasonal growth rate of transplanted white mangroves, in *Proc. Annu. Conf. Restor. Coastal Vegetation Fla.,* Lewis, R. R. and Cole, D. P., Eds., May 14, 1977, Hillsborough Community College, Tampa, Fla., 1978, 77.
66. **Lewis, R. R. and Dunstan, F. M.,** Results of mangrove planting experiments on a spoil island in Tampa Bay, Rep. to the Tampa Port Authority, 1976.
67. **Lewis, R. R., Lewis, C. S., Fehring, W. H., and Rodgers, J. A.,** Coastal habitat mitigation in Tampa Bay, Florida, in *Proc. Mitigation Symp.,* July 16—20, 1979, Colorado State University, Ft. Collins, Colo., Tech. Rep. RM-65, U.S. Department of Agriculture, Ft. Collins, Colo. 1979, 136.
68. **Lewis, R. R.,** unpublished data.
69. **Honey, W.,** Latex Construction Company, Atlanta, Ga., personal communication.
70. **Carlton, J. M. and Moffler, M. D.,** Propagation of mangroves by air-layering, *Environ. Conserv.,* 5(2), 147, 1978.
71. **Steinke, T. D.,** Some factors affecting dispersal and establishment of propagules of *Avicennia marina,* in *Proc. Int. Symp. Biol. Manage. Mangroves,* Walsh, G., Snedaker, S., and Teas, H., Eds., October 8—11, 1974, Honolulu, Hawaii, Institute of Food and Agricultural Sciences, Gainesville, Fla., 1975, 402.
72. **Lewis, R. R.,** unpublished data.
73. **Lewis, R. R.,** unpublished data.
74. **McMillan, C.,** Environmental factors affecting seedling establishment of the black mangrove on the central Texas coast, *Ecology,* 52, 927, 1971.
75. **McMillan, C.,** Salt tolerance of mangroves and submerged aquatic plants, *Ecology of Halophytes,* Reimold, R. J. and Queen, W. H., Eds., Academic Press, New York, 1974, 379.
76. **Connor, D. J.,** Growth of grey mangrove *Avicennia marina* in nutrient culture, *Biotropical,* 1(2), 36, 1969.
77. **Banus, M. D. and Kolemainen, S. E.,** Planting, rooting and growth of red mangrove (*Rhizophora mangle* L.) seedlings: effect on expansion of mangroves in Puerto Rico, in *Proc. Int. Symp. Biol. Manage. Mangroves,* Walsh, G., Snedaker, S., and Teas, H., Eds., October 8—11, 1974, Honolulu, Hawaii, Institute of Food and Agricultural Sciences, Gainesville, Fla., 1975, 370.
78. **Sidhu, S. S.,** Culture and growth of some mangrove species, in *Proc. Int. Symp. Biol. Manage. Mangroves,* Walsh, G., Snedaker, S., and Teas, H., Eds., October 8—11, 1974, Honolulu, Hawaii, Institute of Food and Agricultural Sciences, Gainesville, Fla., 1975, 394.
79. **Stern, W. L. and Voigt, G. K.,** Effect of salt concentration on growth of red mangrove in culture, *Bot. Gaz.,* 121, 36, 1959.
80. **Egler, F. E.,** The dispersal and establishment of red mangrove, *Rhizophora,* in Florida, *Caribb For.,* 9(4), 299, 1948.
81. **Clements, F. E.,** Plant succession: an analysis of the development of vegetation, *Carnegie Inst. Washington, Publ.,* 242, 1, 1917.

82. Chapman, V. J., Mangrove phytosociology, *Trop. Ecol.,* 11(1), 1, 1970.
83. Lewis, R. R. and Lewis, C. S., Colonial bird use and plant succession on dredged material islands in Florida, Vol.II, Patterns of vegetation succession, Environ. Effects Lab., U.S. Army Eng. Waterways Exp. Stn., Vicksburg, Miss., Contract Rep. D-78-14, 1978.
84. Pannier, F., Pannier R. F., and Mizrachi, D., Succession patterns of mangroves established on spoil islands of Lake Maracaibo, Venezuela, in *Prog. 2nd Int. Symp. Biol. Manage. Mangroves,* Papua, New Guinea, 1980, 42.
85. Detweiler, T., Dunstan, F. M., Lewis, R. R., and Fehring, W. K., Patterns of secondary succession in a mangrove community, Tampa Bay, Florida, in *Proc. 2nd Annu. Conf. Restor. Coastal Vegetation Fla.,* Lewis, R. R., Ed., May 17, 1975, Hillsborough Community College, Tampa, Fla., 1976, 52.
86. Crafton, W. M. and Wells, B. W., The old field prisere: an ecological study, *J. Elisha Mitchell Sci. Soc.,* 50, 225, 1934.
87. Lewis, R. R. and Dunstan, F. M., The possible role of *Spartina alterniflora* Loisel in establishment of mangroves in Florida, in *Proc. 2nd Annu. Conf. Restor. Coastal Vegetation Fla.,* Lewis, R. R., Ed., May 17, 1975, Hillsborough Community College, Tampa, Fla., 1976, 82.
88. Lewis, R. R. and Lewis, C. S., Tidal marsh creation on dredged material in Tampa Bay, Florida, in *Proc. 4th Annu. Conf. Restor. Coastal Vegetation Fla.,* Lewis, R. R. and Cole, D. P., Eds., May 14, 1977, Hillsborough Community College, Tampa, Fla., 1978, 45.
89. Niering, W. A. and Egler, F. E., A shrub community of *Viburnum lentago,* stable for twenty five years, *Ecology,* 36, 356, 1955.
90. Snedaker, S., personal communication, 1980.

Chapter 9

SEAGRASS MEADOWS

Ronald C. Phillips

TABLE OF CONTENTS

## I. DESCRIPTION OF SEAGRASS ECOSYSTEM

Seagrasses are vascular plants which produce flowers and are rooted in soft, sandy, or muddy sediments of the nearshore coastal oceanic environment. Horizontal stems, known as rhizomes, occur in the sediment and send erect leafy shoots into the water. The shoots are often dense, with leaves so long that they resemble vast under-water meadows of waving wheat or oats. Approximately 50 species of seagrasses are found throughout the oceans of the world.

The seagrasses are often found in vast underwater meadows of one or more species. Where found with coastal marshes or mangrove systems, the seagrasses form offshore and extend into deeper water. Since the plants grow on a landscape of relatively uniform relief, they create a diversity of habitats and substrates. Owing to their density, an ability to conduct photosynthesis at a relatively high rate, and their rooted presence in the substrate, seagrasses perform a wide assortment of functions in the coastal environment:[1]

1.  Eelgrass, a North Temperate seagrass, has a high rate of growth, producing on the average about 300 to 600 g dry weight per square meter per year, not including root production.
2.  The leaves support large numbers of epiphytic organisms with a total biomass approaching that of the plants themselves. This diversity is possible because of the abundance of oxygen, nutrients, and food provided by the plants.
3.  Although a few organisms may feed directy on the eelgrass and several may graze on the epiphytes, the major food chains are based on eelgrass detritus and its resident microbes.
4.  The organic matter in the detritus and in decaying roots indicates sulfur reduction and maintains an active sulfur cycle.
5.  The roots bind the sediments together, and with the protection afforded by the leaves, surface erosion is reduced, thereby preserving the microbial flora of the sediment and the sediment/water interface. Since seagrass rhizomes form a dense, interlacing mat and the leaves form a dense baffle, the plants are so effective in their hold on the bottom that they persist during tropical hurricanes, despite wave action caused by 150 knot winds.
6.  The leaves return currents and increase sedimentation of organic and inorganic materials around the plants.
7.  Eelgrass absorbs nutrients through the leaves and roots, i.e., nitrogen and phosphorus can be returned to the water column from sediments via seagrasses.

## II. HABITAT AND PRODUCTIVITY VALUES

Many studies have documented the role of seagrasses in forming highly productive ecosystems with a suite of biological and physical functions which transcend the boundaries of the actual seagrass community.[2-5]

Several studies established that seagrass meadows provide nursery and shelter for a great diversity and abundance of both plant and animal species.[4-9] The plants may be unattached and found among the seagrass stalks, they may be attached to rocks or shell fragments on the bottom, or they may be found on the seagrass leaves as epiphytes. The animals may be permanent residents or may be temporary or seasonal transients.

Two kinds of food chains function in seagrass meadows: (1) grazing chains, wherein animals eat the living plant parts and (2) detritus chains, in which plant parts die, decompose, and form particulate matter termed detritus. In the latter chain, owing to the interaction of these organic particles and bacteria, not only are complex, long food chains formed, but also the basis for nutrient cycling and the internalization of organic matter and nutrients produced within the system.

In the North Temperate, eelgrass attracts great numbers of waterfowl, which eat the fresh leaves and seeds. Black brant and American brant geese, whose migration patterns often follow the coastal distribution of eelgrass, are extremely important in the revenues of many states. In the tropics, the green turtle, dugongs, and manatees eat seagrass leaves. All three animals are presently used as food. Many other animals eat seagrasses, but most are not directly used by humans as food.[5]

The tendency of seagrasses to generate detritus rather than enter grazing food chains has been noted.[10] Milne and Milne[11] stated that only 20% of the eelgrass biomass could be directly consumed by fishes and waterfowl and that nutrient material entered food chains primarily from decomposition of detritus. In North Carolina, Thayer et al.[12] reported that eelgrass meadows supported larger populations of invertebrates and fishes than adjacent estuaries. They also found that eelgrass occupied only 17% of the estuarine area, but supplied 64% of the combined total production of phytoplankton, *Spartina*, and eelgrass.

The evidence suggests that microorganisms are the most important consumers of seagrasses and that detritus-feeding invertebrates derive their nourishment primarily by stripping the microorganisms from the plant material as it passes through their guts.[10,13,14] The fecal pellets may be recolonized by microorganisms and the process repeated until all plant material has been utilized.[15] One work[16] demonstrated that many invertebrates can live almost indefinitely on an exclusive diet of bacteria.

Seagrass productivity is very high. Even in the shorter growing season of the North Temperate Zone, eelgrass productivity equals or exceeds that of most cultivated crops (Table 1).[17,18] Representative values for *Thalassia testudinum* in the Caribbean range from 125 to 4000 gC/m²/year and 50 to 960 gC/m²/year for eelgrass in North America. Overall, *Thalassia* productivity is approximately twice that of eelgrass.[17] This could be a function of the length of the growing season. It is assumed that the growing season for *Thalassia* is 250 days, while that for eelgrass is only 120 days.[17]

The epiphytic component of the seagrass ecosystem is very important. Epiphyte productivity was measured at 20% of the mean annual net production of *Thalassia* in Florida (200 gC/m²/year)[19] and about 25% of the annual production of eelgrass in North Carolina.[20] In North Carolina, eelgrass and its epiphytes contribute almost 15% of the total dissolved organic carbon in the estuarine system.[21]

Seagrass meadows are extremely important in the cycling of nutrients. Nitrogen,[22,23] phosphorus,[24] carbon,[25] sulfur,[1] and other nutrients are converted into forms more readily usable by other organisms. These nutrients are taken up by the plants through the roots and pumped into the water mass.

Izembek Lagoon, Alaska, has a surface area of 218 km² with eelgrass covering 116 km². Eelgrass there annually produces 116,000 t (metric tons) of particulate carbon which contains 7400 t of nitrogen, 1660 t of phosphorus, 3.45 t of copper, and 386 t of silica.[26] Only a small fraction of this was recycled within the lagoon. One study[22] found that nitrogen was fixed in the rhizosphere of eelgrass. McRoy and Barsdate[24] reported that eelgrass leaves and roots absorb phosphorus, the major pathway being from the roots to the leaves to the water; but they found that eelgrass could serve as a phosphorus sink in phosphate-rich waters.

Eelgrass appears to maintain an active sulfur cycle.[1] The predominantly reduced environment around the plants allows for anaerobic microbial decomposition of detritus; the sulfides formed create a sink for many toxic metals in the marine environment. The thin oxidized surface layer allows for sulfate accumulation, which is available to the microorganisms involved in the decomposition processes.[1,27]

Table 1
ANNUAL NET PRODUCTION OF SEAGRASSES AND
SELECTED CROP PLANTS (as gC/m²/year)

| Species | Live production | Detritus | Dissolved organic matter |
|---|---|---|---|
| *Zostera marina* (assume growing season of 120 days) | | | |
| Leaves | | | |
| Puget Sound, Washington (41) | 84—840 | | |
| Alaska (42) | 396—456 | | |
| Alaska (43) | 960 | | |
| North Carolina (21) | | 330 | 5.0 |
| North Carolina (20) | 24—204 | | |
| North Caroina (12) | 120 | | |
| New York (44) | 580 | | |
| Rhode Island (45) | 48—348 | | |
| Rhode Island (122) | 3—9 | | |
| Massachusetts (52) | 5 | | |
| Canary Islands (53) | 42 | | |
| Denmark (54) | 240—876 | | |
| Denmark (59) | 108—384 | | |
| France (117) | 422 | | |
| Japan | | | |
| Winter (January; 118) | 7.2 gC/m²/day | | |
| Epiphytes | | | |
| North Carolina (21) | | 73 | 1.5 |
| Massachusetts (51) | 20 | | |
| *Zostera japonica* (55) | 8—10 | | |
| *Phyllospadix scouleri* (assume growing season of 120 days) | | | |
| California (123) | 1440 | | |
| Alaska (124) | 778—1267 | | |
| *Thalassia testudinum* (assume growing season of 250 days) | | | |
| Florida (48) | 500 | | |
| Florida (19) | 900 | | |
| Florida (49) | 425—575 | | |
| Florida (125) | 720—2600 | | |
| Cuba (126) | 58—656 | | |
| Texas (124) | 57—257 | | |
| Bahamas (50) | 1400 | | |
| Barbados (50, 60) | 125—750 | | |
| Jamaica (61) | 475—750 | | |
| Epiphytes | | | |
| Florida (19) | 200 | | |
| *Halodule wrightii* (assume growing season of 120 days in North Carolina; 250 days in Florida) | | | |
| North Carolina (20) | 58—240 | | |
| Florida (128) | 200 | | |
| *Posidonia oceanica* (assume growing season of 120 days) | | | |
| Malta (56) | 240—600 | | |
| *Syringodium filiforme* (assume growing season of 250 days) | | | |
| Florida (62) | 200 | | |
| Florida (128) | 158 | | |

## Table 1 (continued)
## ANNUAL NET PRODUCTION OF SEAGRASSES AND
## SELECTED CROP PLANTS (as gC/m²/year)

| Species | Live production | Detritus | Dissolved organic matter |
|---|---|---|---|
| *S. isoetifolia* (assume growing season of 250 days) | | | |
| Laccadives (57) | 1453 | | |
| *Posidonia sinuosa* | | | |
| Western Australia (58) | 25—282 | | |
| *P. australis* | | | |
| Western Australia (58) | 46—393 | | |
| *Thalassia hemprichii* | | | |
| Palau (63) | 28—281 | | |
| *Enhalus acoroides* | | | |
| Paula (63) | 422—689 | | |
| Papua New Guinea (64) | 118 | | |
| *P. oceanica* | | | |
| Gulf of Naples (120) | 1182 | | |
| Cultivated crops (World averages; 18) | | | |
| Wheat | 344 | | |
| Corn | 412 | | |
| Rice | 497 | | |
| Hay | 420 | | |
| Sugar beets | 765 | | |
| Sugar cane | 1725 | | |

There is an important relationship between seagrass detritus formation and nutrient cycling, both within and across ecosystem boundaries. Nutrients enter seagrass food chains primarily through the transport and decomposition of detritus, despite the direct consumption of about 20% of the biomass by fishes and waterfowl.[11]

Microbial decomposition of detritus is of prime importance in nutrient release and cycling. Many nutrients are released as plant exudates during the growth of the plant. Mann[10] reported that fresh and senescent leaves of eelgrass contain about 20 and 12%, respectively, water-soluble organics. Particulate detritus is poor in essential nutrients, while bacteria contain very high amounts of phosphorus and nitrogen.[28] Bacteria are located as a film around detritus particles. These bacteria absorb nutrients from the water and while acting on the detritus, enrich it with N and P. Mineral nutrients cycle between bacteria and animals, the latter remineralizing the nutrients by digesting the bacteria. For example, bacteria store phosphorus, while animals excrete it. Concurrently, eelgrass plants in the system excrete dissolved organic carbon[21] into the water column which is available to epiphytes on the seagrass blades and to benthic macroalgae and microalgae in the meadows.

Particulate detritus from seagrass blade decomposition may remain in the water column or may settle to the bottom where it may enter an initial phase of oxidation (eelgrass sediments are oxidized in a small surface layer), but soon it enters an anaerobic layer below the surface.[29] Because of this and an abundance of sulfur bacteria, sediments tend to be dominated by the sulfur cycle. Although anaerobic decomposition is slow, it favors the release of mineral nitrogen, phosphorus, and readily assimilable organic constituents. There is additional evidence that anaerobic decomposition, in the presence of sulfate, produces six times more organic material than that occurring by aerobic decomposition.[14] This appears to be related to rich microflora and micro- and macrofauna.[28-32]

Owing to the density and length of leaves, seagrass meadows form a baffle which increases the rate of particle sedimentation, preferentially concentrating finer particle sizes.[33] They do this by the entrapment of water-borne particles by the leaf blades, by the formation and retention of particles produced within the meadows, and by the binding and stabilization of the substrate by the rhizome and root systems. These effects can be local or widespread. Long-term widespread influences of seagrasses on sedimentation include: (1) the carbonate bank along the eastern margin of Shark Bay in Western Australia,[34] (2) the grassbound "mattes" on the Mediterranean coast of France,[35] and (3) the Tavernier and Rodrigues Banks off Florida.[36] Cottam and Munro[37] noted that when eelgrass declined along the North Atlantic coast in the 1950s, the sediments progressed from organic fine silts to unstable sands. Fenchel[27] and Barsdate et al.[26] noted that the fine-textured sediment added stability to the substrate and greater nutrition for eelgrass.

The effect of seagrass meadows on sediment stabilization is well-documented. Sand banks, formerly covered by eelgrass, were lowered by 30 cm or more almost overnight in Salcombe Harbor, Great Britain, after the plants disappeared in 1931.[38] Many species of filter feeding invertebrates, molluscs, and several flatfishes also disappeared. Up to 20 cm of sediment eroded from unvegetated sand banks following a single storm in Chesapeake Bay, while little, if any, sediment disappeared from within an eelgrass meadow.[32] Thomas et al.[39] recorded the effects of winds from Hurricane Donna in 1960 on *Thalassia* in Biscayne Bay in Florida. Winds gusted to 80 mi/hr in the northernmost portion of the bay and exceeded 100 mi/hr in the southern end of the bay for 24 hr before and after the passage of the eye of the storm. Although windrows of leaves were cast up on shore, only light damage was done to the turtlegrass. Only a slight thinning of plants was noted in shallow water. They concluded that the high growth rate of *Thalassia* leaves (up to 25 mm increment per week) would lead to early recovery. They also stated that fresh water from rains during the storm probably contributed the little damage suffered by the plants.

Sediment stability is a function of density and the seagrass species.[33] *Thalassia* (wide, flat blade) is more effective at binding sediment than is *Syringodium* (narrow, terete blade). Observations indicate that near the edge of an eelgrass meadow, there is an increasing density of plants and an increasing amount of sediment removed by turbulent water. Finally Orth[32] found a positive correlation between sediment stability and invertebrate infaunal diversity.

Since seagrass systems trap sediments and organic materials, they may filter effluents.[5] In the "wasting disease" of eelgrass which began in 1931 in the North Atlantic, the value of a *Zostera* meadow in the filtering of raw sewage was established when valuable benthos were poisoned after the plants were removed.[40] There have been discussions as to using seagrass ecosystems as tertiary treatment centers in the U.S., but no work has been done as yet.

## III. HISTORY OF HABITAT LOSS, INCLUDING MODIFICATIONS OF SPECIFIC AREAS

It would be impossible to list all natural and human-induced habitat losses and modifications which affect seagrasses. Only those large-scale losses that have occurred, and that appear to be occurring, will be treated.

### A. The Eelgrass "Wasting Disease" in the North Atlantic

In 1931 a severe decline in *Zostera marina* (eelgrass) began on the east coast of North America and spread to Europe. The "disease" escalated by August 1932 to most locations in the North Atlantic. By 1933, eelgrass was disappearing from Denmark, Sweden, and Norway.[65] By 1933 up to 90 to 99% of all the eelgrass had disappeared from the North Atlantic. Stocks made slight recoveries in the late 1930s and fluctuated in the 1940s. The 1950s showed continued improvements. In 1960, really vigorous growths of eelgrass were observed.[65]

Early explanations for the declines centered around disease organisms: fungi,[66] mycetozoans,[67,68] bacteria,[69] and natural phenomena such as sunspots[70] and warming of the water.[71,72] Rasmussen[65] recently established that the North Atlantic was much warmer during the late 1920s and early 1930s than normal. These higher temperatures probably exceeded the tolerances of extant strains to survive. As they became weakened, they were secondarily invaded by a variety of organisms which hastened their decomposition. The conclusion was that eelgrass growth, development, and survival are primarily dependent on temperature.[65]

## B. Dredging Activities

Large-scale dredging in luxuriant seagrass meadows began in the early 1950s in Boca Ciega Bay near St. Petersburg, Fla.[73] Between 1950 and 1968 dredging and filling activities for real estate reduced turtlegrass area in Boca Ciega Bay by 20%. By the late 1960s, commercial hydraulic clam dredging was occurring in the same bay.[74] Trenches up to 1 m deep were blasted out of the seagrass meadows.

Recently, attention has focused on the long-term damage done to turtlegrass meadows from outboard engine propellors.[75] Tracks cut through the meadows were conspicuous up to 5 years afterward.

A small-scale dredging project begun in the autumn of 1975 in the Aransas Channel near Port Aransas, Tex. resulted in the deposition of 15 to 25 cm of sand and silt over indigenous *Halodule* and transplanted *Halodule* and *Thalassia* for at least 13 months, perhaps as long as 21 months.

Dredging in seagrass meadows for coral sand in Suva, Fiji, left continuous pits 7 to 11 m keep[164].

## C. Damages Caused by Organisms

The sea urchin, *Lytechinus variegatus*, caused massive destruction of fixed *Thalassia-Syringodium-Halodule* meadows off the west coast of Florida[76] over a 26 km distance (5.5 to 9.25 km offshore). Urchin densities at the leading edges of the front averaged 63.6/m², with individuals piling up 2 to 8 deep. Several meters behind the front, densities averaged 5.6/m². The aggregates moved through the grass an average rate of 1.6 m/week. They found that an aggregate 9 m wide could have denuded an area approximately 14 m² in 1 week.[76]

In the lower Chesapeake Bay eelgrass declined rapidly from 1971 to 1974 (up to 36% reduction).[77] By 1975 there was a virtual absence of all eelgrass in the lower Bay area. This was attributed to an influx of cownose rays which were digging for clams.

## D. Heated Water Effluents

The best documented study of the impact of thermal effluents is that from two fossil fuel power plants erected in April 1967 and 1968 at Turkey Point, Biscayne Bay,[78]. Water temperature from the two power plants was elevated as much as 5°C above ambient.

Between 1959 and 1963 three power generating units were turned on in Tampa Bay, Fla. Intake water averaged 16°C in the coldest month and 32°C in the warmest month. Both *Thalassia* and *Halodule* occurred around the intake side of the plant. Discharge water averaged 22°C in the coldest month and 37.5°C in the warmest month. By 1974 only *Halodule* occurred near the discharge side of the plant.[79]

## E. Eutrophication, Heavy Metals, and Various Organics

One of the most luxuriant growths of *Thalassia* and epiphytic algae ever found was recorded just inshore of a Miami sewage plant on Virginia Key, Fla.[80] Nearby in Biscayne Bay, no seagrasses occurred in an area receiving sewage effluents.[81] Only *Halophila decipiens* and *Halodule* persisted within 1 km of the outfall; beyond 1 km, *Thalassia* occurred sparsely. However, significant portions of Hillsborough Bay, Fla. were damaged by dredging and domestic and industrial effluents, particularly phosphorus and domestic and industrial effluents, phosphorus and nitrogen compounds, and suspended solids.[82] Effects noted were heavy growths of phytoplankton and filamentous algae, marked fluctuations in oxygen, and high turbidity. The latter resulted in reduced density and coverage of *Thalassia*. Fish and crustacean catches were low as compared to other areas where seagrasses were extensive. Lewis et al.[83] noted an 80% reduction in seagrasses in Hillsborough Bay in 100 years, ostensibly due to a reduction in water quality (nutrients, sediment loading).

In northwest Florida, Escambia Bay receives the discharges of 23 significant (wastewater flows greater than 378 m³/day or 0.1 million gal/day) municipal-private domestic point sources and 10 significant industrial point sources. Most of the grass beds in Escambia Bay disappeared from 1949 to 1966. A study showed that most of the largest industrial and municipal sources initiated runoffs between 1957 to 1962.[86]

Extensive *Thalassia* was replaced by *Enteromorpha* in Christensted Harbor, St. Croix, U.S. Virgin Islands, when sewage inputs and dredging increased.[84] Over a 17 year period seagrass beds underwent a 66% reduction in this harbor.[85]

In 1974 Peres[165] predicted a total demise of seagrasses along the Mediterranean coast of France owing to sewage pollution.

In Cockburn Sound, Western Australia, there were major losses of seagrasses attributed to nutrient enrichment, particularly nitrogen. These effluents came from two shore industries.[58] In 1979, only 22% of the seagrasses present before the release of the Kwinana Nitrogen Company (KNC) effluents remained. Leaf production was reduced sixfold in 25 years.

In northwest Florida *Thalassia* was predominant in the Econfina River (unpolluted) drainage, but died off in the Fenholloway River (polluted) drainage. The pollution source was kraft mill effluents from a pulp mill. Even though *Syringodium* and *Halophila engelmanni* persisted off the Fenholloway River, biomass was lowered significantly. In the most heavily polluted areas, all seagrasses and algae were gone.[131]

There are no data on any inimical effects of heavy metals on seagrass growth. Studies in Long Island Sound, New York, documented cadmium, manganese, and zinc uptake[88-90] in eelgrass with no apparent effects. Likewise, in Texas bays seagrasses have been shown to concentrate cobalt and manganese (up to ten times the concentration in the sediments), iron[91] and zinc (five to ten times the concentration in the sediments)[92] without apparent damage. However, they may make these excess metals available for movement up the food chain.[80] It is likely that manganese and iron move from *Thalassia* to sea urchins.[93]

Only one study has been done on the effects of herbicides on seagrasses. In an attempt to eliminate eelgrass in favor of oyster growth, an application of 2,4-D in a concentration of 41 to 81 lb/acre resulted in the death of the plants. Five other herbicides had no effect on the eelgrass.[94]

## F. Effects of Oil Pollution

Seagrasses appear to be susceptible to oil pollution.[95] In Puerto Rico, beds of *Thalassia* were adversely impacted by crude oil on the south shore.[96] Plants deteriorated over a period of several months. There was an attendant loss of 3000 m³ of sand from Tamarindo Beach in less than a week, owing to the combination of oil and the fine-grained sediments, facilitating their removal.[80]

The Santa Barbara oil spill in 1969 heavily coated leaves of *Phyllospadix torreyi*. Those killed were in contact with the air.[97] Those plants in 10 cm of water were protected from any damaging effects of the oil. However, when the oil was removed from the intertidal, new leaves were produced from the rhizomes.[98] In England, *Zostera* shed its leaves after oil contamination, but regenerated new blades from protected roots and shoots.[99] Eelgrass in northern France, coated with oil from the *Amoco Cadiz* spill, lost its leaves initially, but produced new ones from persistent rhizomes when the oil was removed.[100] Animal diversity and abundance was severely reduced initially, but fully recovered within a year (except for the filter-feeding Amphipoda).

## IV. IMPACTS ASSOCIATED WITH HABITAT LOSSES

### A. The Eelgrass ''Wasting Disease'' in the North Atlantic

In North America, loss of animal life was extensive after eelgrass died off. American brant populations declined up to 80% by 1934 in the eastern U.S.[101] or changed their migration patterns. In the Netherlands, up to 2/3 of the brant stocks disappeared.[102] There was a reduction in brant from 250,000 to 22,000 at Scott Head Island in England.[103] Soft-shelled and razor clams, lobsters, and mud crabs declined severely when eelgrass disappeared.[104] Cod, flounder, scallops, crabs, and other food animals were also reduced.[11] In the Niantic River, Conn. scallop populations soared in the absence of eelgrass.[105] In other areas scallops are dependent on eelgrass for spawning.[106] In Massachusetts, one third of the faunal species in certain areas disappeared after eelgrass removal.[107]

In England, up to 30 cm of sediment were removed from one harbor within 2 weeks after the plants were gone.[108] In Denmark, drastic declines in fisheries and other animals were observed in North America and were predicted,[109] but did not occur.[65] It is now thought that the rich organic sediment built up by the eelgrass system over the years began to release nutrients into the water mass in the absence of the plants, cushioning an immediate impact on the fisheries. Further, erosion of the silt and sand sediments following the demise of eelgrass removed the possibility of eelgrass returning in several fjords in Denmark.[65] In these fjords, the nature of the soft bottom animal communities were changed as much of the eelgrass organic layer oxidized off; current patterns were changed in many places as new sand bars formed and fluctuated. The premise is that the major effect resulting from the loss of eelgrass in Denmark was a fundamental change in the substrate of the animal communities.[65] Very few of the eelgrass-associated animals disappeared, as they did in North America.[107] In North America, the bottom was left denuded,[107] while in Denmark the resulting uncovered rocks were quickly covered by *Fucus*.[65]

### B. Dredging Activities

Following the extensive dredging done in Boca Ciega Bay, Fla., between 1950 and 1968 when *Thalassia* coverage was reduced by 20%, studies reported that dredged portions contained less than 20% of the plant and animal species formerly recorded from the undisturbed bay.[73,110,111] Water in Boca Ciega Bay became very turbid in many areas in the late 1950s as dredges were operating in seagrass meadows. Microflora epiphytic on *Thalassia* blades trapped much of the suspended silt.[111] Using conservative and incomplete figures, there as an estimated annual loss in fisheries and water sports of approximately $1.4 million/year from the bayfill operations (1400 acres filled). Where hydraulic clam dredges blasted seagrasses out of Boca Ciega Bay, some recovery by *Halodule* occurred, but no regrowth by *Thalassia* or *Syringodium* was ever observed.[74]

In Texas, the deposition of 15 to 25 cm of sand and silt over indigenous and transplanted *Halodule* and *Thalassia* clearly distinguished adaptive tolerances of these two species to sediment loading.[112] *Thalassia* transplants died under the introduced sediment, while *Halodule* persisted. After the sediment washed off, indigenous and transplanted *Halodule* succumbed following an extremely cold winter.[112]

In Lindbergh Bay, St. Thomas, Virgin Islands, increased turbidity caused by dredging resulted in a marked reduction in *Thalassia* and *Syringodium* distribution and abundance. In 1968, before the dredging *Thalassia* could be seen covering the bottom down to 10 m. By 1971 both species were extremely sparse, with few plants found deeper than 2.5 m. In the most turbid parts of the bay, they were limited to 1.5 m in depth.[129] Similar destruction of seagrass beds by sedimentation and turbidity also occurred following dredging in Brewers Bay, St. Thomas.[130]

In Suva, Fiji, where pits were dug 7 to 11 m deep into the seagrass meadows, recolonization was very slow. For the first 2 years, slumping and collapsing of the pit slopes into the pits occurred. After 2 years, bottom stability was often great enough to allow patchy recolonization from surrounding seagrasses (*Syringodium isoetifolium, Halophila minor*[164]).

### C. Damages Caused by Organisms

In the Chesapeake Bay, up to 20 cm of sediment eroded from unvegetated sand banks following a single storm, while little, if any, sediment disappeared from within an eelgrass meadow.[32]

Off Florida, local residents reported that offshore grassflats denuded of leaves by *Lytechinus* became more denuded in time, even losing rhizomes.[76] It was doubtful if early regrowth would occur.

### D. Heated Water Effluents

Following the release of heated water from two fossil fuel power plants at Turkey Point, Biscayne Bay (one opened in April 1967; the other in April 1968), water temperatures near the plants were raised as much as 5°C above ambient. All seagrasses in an area of 9.1 ha (22.5 acres) disappeared off the mouth of the discharge canal, while those in an area of 30 ha (74.1 acres) further out declined. Animal communities associated with these meadows disappeared.[114,115] By January 1976, the inner totally denuded area contained 0 to 10 blades of *Thalassia* per square meter.[116]

### E. Effects of Organics

Following the removal of eelgrass by 2,4-D in New Brunswick, the bottom sediments lost their fine silts and became much more firm.[121]

### F. Effects of Oil Pollution

Following the damage to *Thalassia* by oil spilled in Tamarindo Beach, Puerto Rico, and the increased buoyancy of the oil and sediment mixture, 3000 m³ of sand was washed out in less than a week.[80,96]

## V. SPECIFIC PROJECTS

### A. Rationale

Seagrass transplantation has allowed investigators to explore basic biological problems: (1) interspecific variations, (2) phenotypic plasticity and intraspecific genotypic differentiation as these relate to adaptive tolerances, (3) phenology, and (4) effects of a variety of environmental pollutants on the survival, development, and growth of seagrasses.[132,133,135] Transplantation can also be used to (1) replenish stocks in areas damaged by pollution disease, or excessive grazing,[2,132,133,135,136] (2) compensate for those lost to reclamation programs,[133] and (3) stabilize dredged materials.[2,134,136]

### B. Techniques

The following methods have been used in seagrass transplantation:[132,137]

1.  Nonanchoring methods.
    a.  Plants washed free of sediment; rhizome mat covered with sediment in transplant site. The units may either be broken up into individual shoots or left intact as large mats of leafy shoots.[134,142,146-148]
    b.  Turfs (plants with sediments intact). These are units of seagrass approximately 0.1 m² (1 ft²) in area dug up and removed from a donor site with a shovel. The depth extends below the rhizome mat.[138,139,143,144]
    c.  Plugs (plants with sediment intact placed in hole in substrate). These units are circular or rectangular. Depending on the species, i.e., eelgrass or *Halodule*, the diameter may be as small as 4 in. (10 cm) or 6 in. (15 cm). For *Thalassia* a diameter of 8 in. (20 cm) works best. The depth should be 10 to 15 cm deep.[138,139,143-145]
    d.  Individual shoots in cans placed in sediment. Plants are dug up, washed free of sediment, and then placed in cans buried in sediment.[139]
    e.  Seeds placed in sediment.
2.  Anchoring methods
    a.  Individual leafy shoots fixed via rubber bands to pipes or construction rods. The rods are then placed in shallow trenches dug into sediment and covered over. In this method, a portion of the rhizome is left on the leafy shoot. Anchors may also be bricks, wire mesh, or nails.[132,138,139,141,142]
    b.  Biodegradable mesh paper. Plants woven in between strands of string and paper.[149]
    c.  Plants fixed to concrete rings and cast onto bottom from a boat.[153-155]
    d.  Seedlings placed in sediment using plastic anchors wrapped around plant.[137]
    e.  Seeds woven into yarn tapes. On one occasion seeds were woven into a knitted tube of yarn, using a toy called Knit Magic® (Mattel®, Inc.).[134]
3.  Several persons have used plant growth hormones before transplanting (5 to 10% NAPH; 5% root dip).[132,134,137-140]

## C. Plant Material Source

Most work to date has used seagrass material for transplantation from an indigenous meadow called a donor site.[87,112,132,134,136,137,144,145,150-155] In one study vegetative material was taken from the area being disturbed by dredging[164] (Suva Fiji; *Syringodium isoetifolium*). Thorhaug, using *Thalassia seeds*, initiated nursery-type procedures for holding and developing transplant stock for later use in the field.[137,150,152] Lewis[166] has recently developed an open water rafting nursery technique for *Thalassia* seed stock with dramatic results in transplant success.

## D. Man-power and Cost Analyses

A recently completed project in the Florida Keys was done using carefully computed time-cost analyses. This project used three seagrass species (*Thalassia testudinum*, *Syringodium filiforme*, *Halodule wrightii*) in various states (vegetative plugs, vegetative shoots, seeds) on different spacings (1 m; 2 m), noting time-manpower data for each combination of planting. This project would be the largest of its kind completed to date with such records. These data are not available for publication.

Only four projects are known from the U.S. that list costs of transplanting seagrasses. In North Carolina, eelgrass was placed on dredged materials using plugs ($76,545/acre; based on cost per shoot)[149] and 5 to 35 vegetative shoots woven through E-Z® fabric mesh to obtain complete bottom cover in about 1980 days ($2,179 to $2,660/acre; planted in water less than 3 ft deep).[149] In Long Island Sound, eelgrass sprigs were planted by hand in the substrate (3 to 4 vegetative shoots on the same rhizome; 1 ft spacings; planted in less than 4 ft deep; 1655 man-hours; $12,775/acre).[134] In San Diego, eelgrass plugs were planted at variable costs depending on the spacings ($25/ft²; 19,400 plugs per acre needed for 18 in. spacing = $242,500; 4840 plugs per acre needed for 36 in. spacing = $60,500; 1 man-hour needed to obtain, move, and plant 1.08 ft² of eelgrass).[156] In Florida, *Thalassia* seedlings were planted at variable costs depending on the desired time to get a desired cover ($33,385/acre to get a cover of 3000 blades per square meter in 2.5 years; $111,286 to get a cover of 1000 blades per square meter in 0.8 year; $222,572 to get a cover of 2000 blades per square meter in 0.8 year; these costs do not include transportation costs, depreciation costs, or administrative overhead).[137] Caution should be taken in using these estimates. They reflect the year the work was done, the hourly wages of the workers, and the experience of the workers. The costs have been developed by small-scale field intensive research rather than by large-scale projects. Therefore, the cost of transplanting can be extremely variable.

In Great Britain, one pilot study using *Zostera noltii* plugs estimated a total of £927/ha or £380/acre using prisoners and students as labor (capital equipment not included; 1973 costs).[133] If only student labor was used, the costs increased 22%.

### E. Success or Failure
This section will describe the various transplant projects that have been done on a world-wide regional basis. These projects include both small-scale experimental efforts as well as larger-scale applied ones, It is certain that others will become known as this is written.

### 1. Temperate North America (Table 2)
The species used is *Zostera marina* (eelgrass). At the extremes of its abundance on the Pacific coast of North America, the species does not tolerate the proximity to iron anchors or the absence of its native sediment when planting. For best results in these areas transplants should be done using plugs. Transplants using seeds of eelgrass have minimal or no success. Seeds of eelgrass have extremely small germination rates in the field at ambient salinities[41] and the seedlings suffer almost total mortality after germination.[41] Also, transplants that have been done using seeds have all failed.[41,134,146-148,157,158] In the middle portion of its range on the Pacific coast, eelgrass survives and expands quickly using virtually all anchoring and nonanchoring methods of vegetative material.

### 2. Japan (Table 3)
Only one group has attempted to transplant seagrasses in Japan.[159-161] This group used seeds of eelgrass held close to the bottom with fine mesh cloth (leaves projecting up through slits in the cloth), seedlings placed in the center of erect bamboo sticks which were sunken into and filled with sediment, and leafy shoots fixed to bamboo sticks and placed in trenches dug into the sediment.

### 3. Great Britain (Table 4)
It appears that seagrass transplantation was attempted in Norfolk and Suffolk, Great Britain.[133] Initial results showed promise, but nothing extensive seems to have developed.

Table 2

## SEAGRASS TRANSPLANTATION IN TEMPERATE NORTH AMERICA

| Species | Propagules | Anchoring method | Location | Chemical additive | Success | Substrate | | Ref. |
|---|---|---|---|---|---|---|---|---|
| | | | | | | Dredged material | Native | |
| Zostera marina L. | Vegetative shoots (sediment washed free) | None | Mass. | None | 0 | | X | 146 |
| | | None | Mass., R.I., Conn., Long Island, N.Y., N.J., Del., Md., Va., N.C. | None | 0 (Plants from Mass. and Pacific Coast) | | X | 147,148 |
| | | None | Great South Bay, Long Island, N.Y. | NAPH and fertilizer in some | Single shoots — 36% (2 months); bunch of 3—4 shoots — 80% (4 months) (no difference when used NAPH or fertilizer) | X X | | 134 |
| | | None | R.I. | None | Not applicable | | X | 151 |
| | | Biodegradable mesh paper | Beaufort, N.C. | None | 91% | | X | 149 |
| | | Wire mesh | San Diego, Calif. | None | Moderate | | X | 167 |
| | | None | San Diego, Calif. | None | Moderate | | X | 167 |
| | | Iron rods | Puget Sound, Wash. | None | 100% | | X | 41 |
| | | Iron rods | Puget Sound, Wash. | None | 67% | | X | 168 |
| | | Nails | Puget Sound, Wash. | None | 50% | | X | 168 |
| | | None | Puget Sound, Wash. | None | 100% | | X | 168 |

## Table 2 (continued)
## SEAGRASS TRANSPLANTATION IN TEMPERATE NORTH AMERICA

| Species | Propagules | Anchoring method | Location | Chemical additive | Success | Substrate | | Ref. |
|---|---|---|---|---|---|---|---|---|
| | | | | | | Dredged material | Native | |
| | Propagules | Nails | Adak, Aleutians, Alaska | None | Very high | | X | 169 |
| | | Nails | Isembek Lagoon, Alaska | None | 50% | | X | 168 |
| | | Iron rods | Isembek Lagoon, Alaska | None | 0 | | X | 168 |
| | | None | Isembek Lagoon, Alaska | None | 100% | | X | 168 |
| | Plugs, turfs (plants with native sediments intact) | | Mass. | None | Some | | X | 157 |
| | | | Long Island, N.Y. | None | 60% (where light was adequate) | | X | 94 |
| | | | Great South Bay, Long Island, N.Y. | | 88% (4 months) | X | | 134 |
| | | | San Diego, Calif. | None | Moderate | | X | 167 |
| | | | San Diego, Calif. | None | 30% | X | | 156 |
| | | | Puget Sound, Wash. | None | 69% | | X | 168 |
| | | | Isembek Lagoon, Alaska | None | 50% (from 1975—1977) | 100% | X | 168 |
| | Seedlings | None | Mass. to N.C. | None | Probably 0 | | X | 146—148, 157, 158 |
| | | Cotton mesh | Great South Bay, Long Island, N.Y. | None | Undetermined (seed germination 0—30% in field; seedling establishment not determined) | X | | 134 |
| | | None | Isembek Lagoon, Alaska | None | 0 | | X | 168 |

187

## Table 3
## SEAGRASS TRANSPLANTATION IN JAPAN

| Species | Propagules | Anchoring method | Chemical additive | Success | Dredged material | Native | Ref. |
|---|---|---|---|---|---|---|---|
| | | | | | \<Substrate\> | | |
| Zostera marina L. | Seeds | Fine net cloth; bamboo sticks | None | 2—5% | | X | 159 |
| | Seeds | Fine net cloth | None | 1.5% | | X | 160 |
| | Vegetative shoots | None | None | 0 | | X | 160 |
| | Seeds | Fine net cloth | None | 6—7% | | X | 161 |
| | Vegetative shoots | Bamboo sticks | None | 100% | | X | 161 |

## Table 4
## SEAGRASS TRANSPLANTATION IN GREAT BRITAIN

| Species | Propagules | Anchoring method | Location | Additive | Success | Dredged material | Native | Ref. |
|---|---|---|---|---|---|---|---|---|
| Zostera noltii Hornem. | Plugs | | Norfolk, Suffolk | None | 35% after 2 years | | X | 133 |

## Table 5
## SEAGRASS TRANSPLANTATION TO FRANCE

| Species | Propagules | Anchoring method | Location | Additive | Success | Dredged material | Native | Ref. |
|---|---|---|---|---|---|---|---|---|
| Posidonia oceanica (L.) Delile | Vegetative shoots | Concrete rings | Marseille | None | Almost 100 | | X | 86, 113, 127, 153—155 |

### 4. France (Table 5)

One project near Marseille used vegetative shoots of *Posidonia oceanica* fixed in the center of a donut-shaped ring of concrete. The plants were fixed to the anchors in the boat, which could then be cast onto the bottom from the boat. Success of transplantation was remarkably high. The method was tested over a large area, and showed that large-scale projects were feasible.[86,113,127,153-155]

### 5. Tropical-Subtropical North America (Table 6)

A large number of projects have been performed, using *Thalassia testudinum, Halodule wrightii,* and to a limited extent, *Syringodium filiforme.* Most of this work was done in Florida.

A project, recently completed for the Florida State Department of Transportation in the Florida Keys, used *Thalassia, Syringodium,* and *Halodule* plugs and individual vegetative shoots and seedlings (1979 to 1981). This material was planted on 1 and 2 m spacings. All plantings were monitored at least quarterly over 2 years. The project also monitored Eh, pH, and a series of nutrients in the water column and sediments of the transplants. Monitoring of the plants and chemistry was done at least quarterly over a period of 2 years. The project is probably the most comprehensive transplant effort done to date.

Work done in Biscayne Bay, using *Thalassia* seeds and seedlings, used culture and nursery techniques including antibiotics to kill bacteria and fungi which would weaken the seedling stocks.[152] All transplants of *Thalassia* and *Halodule* using anchors of iron have resulted in rapid disintegration of the plants.

### 6. Fiji (Table 7)

Dr. Nicholas Penn, University College of Swansea, Swansea, U.K., conducted transplants of *Syringodium isoetifolium* on Suva. The material used was plugs and planting was done at a depth of 5 to 9 m.[164] Artificial reefs, using car bodies, concrete and sewage culverts, and other debris, were built near the transplants to break up currents. There are no results noted at the present.

### 7. Australia (Table 8)

Three projects have been carried out in Australia. One project was located in Cockburn Sound, just south of Perth, Western Australia.[58] Seedlings of both *Posidonia sinuosa* and *P. australis* were used. The seedlings were collected in the field and merely moved to the transplant site. No anchors were used in transplanting.

In Mortons Bay, New South Wales, plugs of *Zostera capricorni* were planted. In Botany Bay, New South Wales, near Sydney, turfs of *Z. capricorni* and *P. australis* were planted. No results were reported.[87] In Illawarra Lake, New South Wales, plugs, turfs, and vegetative shoots of *Zostera capricorni* were used with some success.[162,163]

## VI. RESEARCH NEEDS

A variety of studies are needed on seagrasses to insure the continued vitality and presence of the system in as undamaged a state as possible.[3] These studies should be conducted not only in mildly, or heavily, impacted areas (these areas should give data on the ability of the plants to adapt), but it is imperative that studies be continued in "sanctuaries" or control areas, where the data are indicative of the inherent genetic, morphological, and physiological capabilities of the plants. Seagrasses have been found to be dynamic entities, with a suite of adaptive properties in a changing environment.

Studies are needed to establish improved means of assessing the consequences of

## Table 6
## SEAGRASS TRANSPLANTATION IN TROPICAL — SUBTROPICAL NORTH AMERICA

| Species | Propagules | Anchoring method | Location | Chemical additive | Success | Dredged material | Native | Ref. |
|---|---|---|---|---|---|---|---|---|
| *Thalassia testudinum* Banks ex König | Vegetative shoots (sediment washed free) | Iron Rods | Port Aransas, Tex. | 10% NAPH | 0 | | X | 132 |
| | | | | None | 0 | | | |
| | | Nails | Port Aransas, Tex. | None | 0 | | X | 132 |
| | | Wire mesh | Port Aransas, Tex. | None | 0 | X | | 141 |
| | | Iron rods | Mississippi Sound, Miss. | None | Limited (6 months) | X | | 142 |
| | | Concrete blocks, wire mesh | Mississippi Sound, Miss. | None | 0 | X | | |
| | | | | | 4% | | X | 142 |
| | | None | Mississippi Sound, Miss. | None | 0 | X | X | 142 |
| | | Iron rods | Tampa Bay, Fla. | 10% | 0—100% (total of 30 short shoots tested) | | X | 139 |
| | | | | None | 0—16.7% (total of 30 shoots tested) | | X | |
| | | Iron rods, plastic bags | Tampa, Fla. | 5% NAPH Root Dip | 0—20% | | X | 138 |
| | | | | | 20—40% | | X | |
| | | Concrete rings | Florida Keys | None | 0 | | X | 168, 170 |
| | | | Port Aransas, Tex. | None | 73% (until sediment loading and cold killed them) | | X | 112 |
| | Plugs, turfs, (plants with native sediments intact) | | Port Aransas, Tex. | None | 0 | X | | 141 |
| | | | Near Tallahassee, Fla. | None | Limited | | X | 168 |
| | | | | | 89% | | X | |
| | | | Tampa Bay, Fla. | None | None | | X | 143 |

Table 6 (continued)
SEAGRASS TRANSPLANTATION IN TROPICAL — SUBTROPICAL NORTH AMERICA

| Species | Propagules | Anchoring method | Location | Chemical additive | Success | Substrate | | Ref. |
|---|---|---|---|---|---|---|---|---|
| | | | | | | Dredged material | Native | |
| | | | Tampa Bay, Fla. | None | 15% | X | | 139 |
| | | | Tampa Bay, Fla. | None 5% NAPH 5% Root Dip | 40%; 0—100% (hormones had no effect) (highest when done in winter, i.e., water less than 21°C) | | X | 138 |
| | | | Tampa Bay, Fla. | None | None (test for effect of thermal effluents from power plant) | | X | 79 |
| | | | Escambia Bay, Fla. | None | | | X | 47, 144 |
| | | | Florida Keys | None | 90% (2 m centers) | | | 168, 170 |
| | | | Florida Keys | None | 98% (1 m centers) | | | 168, 171 |
| | Seedlings | None | Near Tallahassee, Fla. | None | 0 (45 seedlings) | | X | 137 |
| | | Plastic | South Biscayne Bay, Fla. | 10% NAPH | 80% | | X | 137 |
| | | Plastic peat pots | North Biscayne Bay, Fla. | 10% NAPH | 5—53% | | X | 140 |
| | | None | Florida Keys | None | 0 (2 m centers); 6% (1 m centers); 18% (⅓ m centers) | | X X X | 168, 170 |
| | | Plastic tags | Florida Keys | | 27% (raised in lab for 6 months) | | | 168, 170 |
| *Halodule wrightii* | Vegetative shoots | Iron rods | Port Aransas, | 10% | 0 | | X | 132 |

191

Ashers.

| Transplant unit | Anchor method | Location | NAPH | Survival | | | Ref. |
|---|---|---|---|---|---|---|---|
| (sediment washed free) | Wire mesh | Tex. | NAPH None | 0 | | X | 141 |
| | | Port Aransas, Tex. | None | 0 | X | X | 142 |
| | Iron rods, concrete blocks, wire mesh | Mississippi Sound, Miss. | None | Limited | | | |
| | | | None | 0 | | | |
| | | | | 13% | | | |
| | No anchors, but placed in sediment in aquarium | Beaufort, N.C. | None | 0 | | X | 149 |
| | | Escambia Bay, Fla. | | | | | 144 |
| | | Indian River, Fla. | 0.05% NAPH | 28% | | | 46 |
| | | | 0.1% NAPH | 45% | | | |
| | | | 0.5% NAPH | 73% | | | |
| | | | 1.0% NAPH | 75% | | | |
| | Concrete rings | Florida Keys | None | 0 (2 m centers) | | X | 168, 170 |
| | | | | 0 (1 m centers) | | | |
| Plugs, turfs (plants with native sediments intact) | | Port Aransas, Tex. | None | 58% (until sediment loading and cold set in) | X | X | 112 |
| | | Port Aransas, Tex. | None | 0 | | X | 141 |
| | | Escambia Bay, Fla. | None | Limited | | X | 144 |
| | | | | 39% | | | |
| | | St. Joe Bay, Fla. | None | 13% overall (after 1 year) (ultimately eroded away by surge from hurricanes) | X | | 145 |
| | | Near Tallahassee, Fla. | None | 50% | | X | 168 |

Table 6 (continued)
SEAGRASS TRANSPLANTATION IN TROPICAL — SUBTROPICAL NORTH AMERICA

| Species | Propagules | Anchoring method | Location | Chemical additive | Success | Substrate Dredged material | Native | Ref. |
|---|---|---|---|---|---|---|---|---|
| | Vegetative shoots (sediment washed free) | | Escambia Bay, Fla. | None | 37% (after 1 year) < 10% where 3 inches of sand covered them | | | 47, 144 |
| | | | Florida Keys | None | 0 (2 m centers) | | X | 168, 170 |
| | | | | | 0 (1 m centers) | | X | |
| | | | | | 61% (1 m centers; second planting using stocks from deeper water) | | X | |
| *Syringodium filiforme* Kutz | Vegetative shoots (sediment washed free) | Iron rods, concrete blocks, wire mesh | Mississippi Sound, Miss. | None | 0 | X | X | 142 |
| | | Concrete rings | Florida Keys | None | 0 (2 m centers) | | X | 168, 170 |
| | | | | | 0 (1 m centers) | | | |
| | Plugs | | Tampa, Fla. | None | 100% | | X | 138 |
| | | | Florida Keys | None | 0 (2 m centers) | | | 168, 170 |
| | | | | | 61% (1 m centers; those remaining expanded over 75% of a 49 m² plot) | | | |
| *Ruppia maritima* L. | Vegetative shoots (sediment washed free) | Wire mesh | Port Aransas, Tex. | None | 0 | | X | 141 |
| | Plugs | | Port Aransas, Tex. | None | 0 | | X | 141 |

## Table 7
### SEAGRASS TRANSPLANTATION IN FIJI

| Species | Propagules | Anchoring method | Location | Chemical additive | Success | Substrate Dredged material | Native | Ref. |
|---------|-----------|------------------|----------|-------------------|---------|----------------------------|--------|------|
| *Syringodium isoetifolium* (Aschers.) Dandy | Plugs | | Suva | None | Not reported | | X | 164 |

environmental change.[3] More studies are needed on nutrient, trace metal, and heavy metal cycling between seagrasses and their sediments. More work is needed on seagrass productivity and the factors that reduce and increase it; harvesting procedures; nutritional analyses; the relative roles of seeds and vegetative growth in maintenance of established meadows; and tolerances of various seagrasses from differing locations to temperature, salinity, depth, sediment, and current differences.

## ACKNOWLEDGMENTS

This material is based upon research supported in part by the National Science Foundation, International Decade of Ocean Exploration, Living Resources Program, under Grants OCE74-24358, OCE-7684259, and OCE77-25559. Seattle Pacific University, Washington, also aided with grants from the Institute for Research.

**Table 8**
**SEAGRASS TRANSPLANTATION IN AUSTRALIA**

| Species | Propagules | Anchoring method | Location | Chemical additive | Success | Dredged material | Native | Ref. |
|---|---|---|---|---|---|---|---|---|
| *Zostera capricorni* Aschers | Turfs | | Moretons Bay, New South Wales | None | None reported | | X | 87 |
| *Z. capricorni* Aschers | Turfs | | Botany Bay, New South Wales near Sydney | None | None reported | | X | 87 |
| *Posidonia australis* Hooker | Turfs | | Botany Bay, New South Wales, near Sydney | None | None reported | | X | 87 |
| *P. australis* Hooker | Seedlings | None | Cockburn Sound, near Perth | None | 70% (control site) | | X | 58 |
| *P. sinuosa* Cambridge et Kuo | Seedlings | None | Cockburn Sound, near Perth | None | 56% (control site) | | X | 58 |
| *Z. capricorni* Aschers | Plugs | | Illawarra Lake, New South Wales | None | 100% (expanded coverage 10 times in 3 years) | | X | 162 |
| *Z. capricorni* Aschers | Vegetative shoots (sediment washed free) | None | Illawarra Lake, New South Wales | None | 0 | | X | 162 |
| *Z. capricorni* Aschers | Turfs | | Illawarra Lake, New South Wales | None | Initiated in October 1975; 136 planted — all survived for 3 months; after that 128 lived and spread 2 to 3 times the original size | X | | 163 |

| *Z. capricorni* Ashers | Turfs | Illawarra Lake, New South Wales | None | Initiated in June 1975; 40 planted; in 3 months only 14 survived but these enlarged 2 to 3 times original size; after 1 year all merged to form a continuous belt 12 m long and 6 m wide | X |

# REFERENCES

1. **Wood, E. J. F., Odum, W. E., and Zieman, J. C.,** *Influence of Seagrasses on the Productivity of Coastal Lagoons,* Laguna Costeras, un simposio, Mem. Simp. Int. Lagunas Costeras, November 28—30, 1967, Mexico, D. F., 1969, 495.
2. **Phillips, R. C.,** Seagrasses and the coastal marine environment, *Oceanus,* 21, 30, 1978.
3. **Phillips, R. C.,** Role of seagrasses in estuarine systems, in Proc. Gulf of Mexico Coastal Ecosystems Workshop, Fore, P. L. and Peterson, R. D., Eds., U.S. Fish and Wildlife Service, Albuquerque, N.M., 1980, 67.
4. **McRoy, C. P.,** Seagrass ecosystems: recommendations for research programs, in *Proc. Int. Seagrass Workshop,* McRoy, C. P., Ed., Leiden, The Netherlands, October 22—26, 1973, Natl. Sci. Foundation, 1973.
5. **McRoy, C. P. and Helfferich, C.,** Applied aspects of seagrasses, in *Handbook of Seagrass Biology: An Ecosystem Perspective,* Phillips, R. C. and McRoy, C. P., Eds., Garland STPM Press, New York, 1980, 297.
6. **Thayer, G. W. and Phillips, R. C.,** Importance of eelgrass beds in Puget Sound, *Mar. Fish. Rev.,* 39, 18, 1977.
7. **Blegvad, H.,** Food and conditions of nourishment among the communities of invertebrate animals found on or in the sea bottom in Danish waters, *Rep. Danish Biol. Stn.,* 22, 41, 1914.
8. **Blegvad, H.,** On the food of the fish in the Danish waters within the skaw, *Rep. Danish Biol. Stn.,* 24, 17, 1916.
9. **Petersen, C. G. J. and Boysen-Jensen, P.,** Valuation of the sea. I. Animal life of the sea bottom, its food and quantity, *Rep. Danish Biol. Stn.,* 20, 1, 1911.
10. **Mann, K. H.,** Macrophyte production and detritus food chains in coastal waters, *Mem. Ist. Ital. Idrobiol.,* 29, 353, 1972.
11. **Milne, L. J. and Milne, M. J.,** The eelgrass catastrophe, *Sci. Am.,* 184, 52, 1951.
12. **Thayer, G. W., Wolfe, D. A., and Williams, R. B.,** The impact of man on seagrass ecosystems, *Am. Sci.,* 63, 288, 1975.
13. **Fenchel, T.,** Studies on the decomposition of organic detritus derived from the turtlegrass, *Thalassia testudinum, Limnol. Oceanogr.,* 15, 14, 1970.
14. **Fenchel, T.,** Aspects of decomposer food chains in marine benthos, *Verh. Deutsch. Zool. Ges.,* 14, 14, 1972.
15. **Harrison, P. G.,** Decomposition of macrophyte detritus in seawater: effects of grazing by amphipods, *Oikos,* 28, 165, 1977.
16. **Zobell, C. E. and Feltham, C. B.,** Bacteria as food for certain marine invertebrates, *J. Mar. Res.,* 4, 312, 1938.
17. **McRoy, C. P. and McMillan, C.,** Production ecology and physiology of seagrasses, in *Seagrass Ecosystems: A Scientific Perspective,* McRoy, C. P. and Helfferich, C., Eds., Marcel Dekker, New York, 1977, 53.
18. **Odum, E. P.,** *Fundamentals of Ecology,* Saunders, Philadelphia, 1959.
19. **Jones, J. A.,** Primary Productivity by the Tropical Marine Turtlegrass, *Thalassia testudinum* König, and its Epiphytes, Ph.D. thesis, University of Miami, Coral Gables, Fla., 1968.
20. **Dillon, C. R.,** A Comparative Study of the Primary Productivity of Estuarine Phytoplankton and Macrobenthic Plants, Ph.D. thesis, University of North Carolina, Chapel Hill, 1971.
21. **Penhale, P. A. and Smith, W. O.,** Excretion of dissolved organic carbon by eelgrass (*Zostera marina*) and its epiphytes, *Limnol. Oceanogr.,* 22, 400, 1977.
22. **Patriquin, D. G. and Knowles, R.,** Nitrogen fixation in the rhizosphere of marine angiosperms, *Mar. Biol.,* 16, 49, 1972.
23. **McRoy, C. P., Barsdate, R. J., and Nebert, M.,** Phosphorus cycling in n eelgrass (*Zostera marina* L.) ecosystem, *Limnol. Oceanogr.,* 17, 58, 1972.
24. **McRoy, C. P. and Barsdate, R. J.,** Phosphate absorption in eelgrass, *Limnol. Oceanogr.,* 15, 6, 1970.
25. **McRoy, C. P.,** Seagrass productivity: carbon uptake experiments in eelgrass, *Zostera marina, Aquaculture,* 4, 131, 1974.
26. **Barsdate, R. J., Nebert, M., and McRoy, C. P.,** Lagoon contributions to sediments and water of the Bering Sea, in *Oceanography of the Bering Sea,* Hood, D. W. and Kelly, E. J., Eds., University of Alaska, Fairbanks, Occas. Publ. 2, 1974, 553.
27. **Fenchel, T.,** Aspects of the decomposition of seagrasses, in *Proc. Int. Seagrass Workshop,* McRoy, C. P., Ed., Leiden, The Netherlands, October 22—26, 1973, Natl. Sci. Foundation, 1973.
28. **Fenchel, T.,** Aspects of the decomposition of seagrasses, in *Seagrass Ecosystems: A Scientific Perspective,* McRoy, C. P. and Helfferich, C., Eds., Marcel Dekker, New York, 1977, 123.

29. Fenchel, T., The ecology of marine microbenthos IV, structure and function of the benthic ecosystem, its chemical and physical factors and the microfauna communities with special reference to the ciliated protozoa, *Ophelia,* 6, 1, 1969.
30. Fenchel, T. and Riedl, R. J., The sulfide system: a new biotic community underneath the oxidized layer of marine sand bottoms, *Mar. Biol.,* 7, 255, 1971.
31. Orth, R. J., Benthic in fauna of eelgrass, *Zostera marina,* beds, *Chesapeake Sci.,* 14, 258, 1973.
32. Orth, R. J., The importance of sediment stability in seagrass communities, in *Ecology of Marine Benthos,* Coull, B. C., Ed., University of South Carolina Press, Columbia, 1977, 122.
33. Schubel, J. R., Some comments on seagrasses and sedimentary processes, in *Int. Seagrass Workshop,* McRoy, C. P., Ed., Leiden, The Netherlands, October 22—26, 1973, Natl. Sci. Foundation, 1973.
34. Davies, G. R., Carbonate bank sedimentation; eastern Shark Bay, Western Australia, in *Carbonate Sedimentation and Environments, Shark Bay, Western Australia,* Logan, B. W., Davies, G. R., Read, J. F., and Cebulski, D. E., Eds., Am. Assoc. Pet. Geol., Mem. 13, Tulsa, Okla., 1970, 85.
35. Molinier, R. and Picard, J., Recherches sur les herbiers de phanerogames marines du littoral mediterranean francais, *Ann. Inst. Oceanogr.,* 27, 157, 1952.
36. Baars, D. L., Petrology of carbonate rocks, in *Shelf Carbonates of the Paradox Basin,* Four Corners Geological Society Symp., 4th Field Conf., 1963, 101.
37. Cottam, C. and Munro, A. D., Eelgrass status and environmental relations, *J. Wildl. Manage.,* 8, 449, 1954.
38. Wilson, D. P., The decline of *Zostera marina* L. at Salcombe and its effects on the shore, *J. Mar. Biol. Assoc. U.K.,* 29, 295, 1949.
39. Thomas, L. P., Moore, D. R., and Work, R. C., Effects of Hurricane Donna on the turtle grass beds of Biscayne Bay, Florida, *Bull. Mar. Sci.,* 2, 191, 1961.
40. McConnaughey, B. H., *Introduction to Marine Biology,* C. V. Mosby, St. Louis, 1974.
41. Phillips, R. C., Ecological Life History of *Zostera marina* L. (Eelgrass) in Puget Sound, Washington, Ph.D. thesis, University of Washington, Seattle, 1972.
42. McRoy, C. P., Standing stocks and other features of eelgrass (*Zostera marina*) populations on the coast of Alaska, *J. Fish. Res. Board Can.,* 27, 1811, 1970.
43. McRoy, C. P., The standing Stock and Ecology of Eelgrass (*Zostera Marina* L.) in Izembek Lagoon, Alaska, M.S. thesis, University of Washington, Seattle, 1966.
44. Burkholder, P. R. and Doheny, T. E., The biology of eelgrass with special reference to Hempstead and South Oyster Bays, Nassau County, Long Island, New York, Lamont Geol. Lab., Palisades, New York, 1968.
45. Conover, J. T., The importance of natural diffusion gradients and transport of substances related to benthic plant metabolism, *Bot. Mar.,* 6, 1, 1968.
46. Zimmerman, C. F., French, T. D., and Montgomery, J. R., Transplanting and survival of the seagrass *Halodule wrightii* under controlled conditions, *Northeast Gulf Sci.,* 4, 131, 1981.
47. Olinger, L. W., Rogers, R. G., Fore, P. L., Todd, R. L., Mullins, B. L., Bisterfeld, F. T., and Wise, L. A., Environmental and recovery studies of Escambia Bay and the Pensacola Bay System, Florida, U.S. Environmental Protection Agency, Region IV, 1975.
48. Iverson, R. L., Harris, R., Bittaker, H. F., and DiDomenico, D., Significance of seagrasses as a source of organic carbon for the eastern Gulf of Mexico coastal zone, Am. Soc. Limnol. Oceanogr., Winter Meetings, January 2—5, 1979.
49. Zieman, J. C., A Study of the Growth and Decomposition of the Seagrass, *Thalassia testudinum,* M.S. thesis, University of Miami, Coral Gables, Fla., 1968.
50. Patriquin, D. G., Estimation of growth rate, production, and age of the marine angiosperm *Thalassia testudinum* König, *Caribb. J. Sci.,* 13, 111, 1973.
51. Marshall, N., Food transfer through the lower trophic levels in the benthic environment, in *Marine Food Chains,* Steele, J. H., Ed., University of California Press, Berkeley, 1970, 58.
52. Conover, J. T., Seasonal growth of benthic marine plants as related to environmental factors in an estuary, *Publ. Inst. Mar. Sci. Univ. Tex.,* 5, 97, 1958.
53. Johnston, C. S., The ecological distribution and primary production of macrophytic marine algae in the Eastern Canaries, *Int. Rev. Gestamen Hydrobiol.,* 54, 473, 1969.
54. Petersen, C. G. J., Om Baendel tangens (*Zostera marina*) Aarsproduktion i de danske Farvande, in *Mindeskriff i Anledning of Hundred aaret for Japetus Steenstrups fødsel.,* Jungersen, F. E. and Warming, J. E. B., Eds., B. Lunos Bogtrykkeri, Copenhagen, 1914.
55. Ogata, E. and Matsui, T., Photosynthesis in several marine plants of Japan as affected by salinity, drying-and pH with attention to their growth habitats, *Bot. Mar.,* 8, 199, 1965.
56. Drew, E. A., Botany, in *Underwater Science,* Woods, J. D. and Lythgoe, J. N., Eds., Oxford University Press, London, 1971, chap. 6.
57. Qasim, S. Z. and Bhattathiri, P. M. A., Primary production of a seagrass bed on Kavaratti Atoll (Laccadives), *Hydrobiologia,* 38, 29, 1971.

58. **Cambridge, M. L.,** Technical report on seagrass, Cockburn Sound Environmental Study, Department of Conservation and Environment, University of Western Australia, Rep. No. 7, 1979.

59. **Sand - Jensen, K.,** Biomass, net production and growth dynamics in an eelgrass (*Zostera marina* L.) population in Vellerup Vig, Denmark, *Ophelia,* 14, 185, 1975.

60. **Patriquin, D. G.,** The origin of nitrogen and phosphorus for growth of the marine angiosperm *Thalassia testudinum, Mar. Biol.,* 15, 35, 1972.

61. **Greenway, M.,** The effects of cropping on tbe growth of *Thalassia testudinum* (Konig) in Jamaica, *Aquaculture,* 4, 199, 1974.

62. **Zieman, J. C. and Wetzel, R. G.,** Productivity in seagrasses: methods and rates in *Handbook of Seagrass Biology: An Ecosystem Perspective,* Phillips, R. C. and McRoy, C. P., Eds., Garland STPM Press, New York, 1980, chap. 7.

63. **Ogden, J. C. and Ogden, N. B.,** A preliminary study of two representative seagrass communities in Palau (Belau), Western Caroline Islands, *Aquat. Bot.,* 12, 229, 1982.

64. **Johnstone, I. M.,** Papua New Guinea seagrasses and aspects of the biology and growth of *Enhalus acoroides* (L.F.) Royle, *Aquat. Bot.,* 7, 197, 1979.

65. **Rasmussen, E.,** The wasting disease of eelgrass (*Zostera marina*) and its effects on environmental factors and fauna, in *Seagrass Ecosystems: A Scientific Perspective,* McRoy, C. P. and Helfferich, C., Eds., Marcel Dekker, New York, 1977, 1.

66. **Petersen, H. E.,** Wasting disease of eelgrass (*Zostera marina*), *Nature (London),* 132, 1004, 1933.

67. **Renn, C. E.,** The wasting disease of *Zostera marina, Biol. Bull.,* 70, 148, 1936.

68. **Young, E. L.,** Studies on *Labyrinthula,* the etiologic agent of the wasting disease of eelgrass, *Am. J. Bot.,* 30, 596, 1943.

69. **Lami, R.,** Travaux recents sur la maladiedes Zosteres, *Rev. Bot., Appl. Agric. Trop.,* 15, 263, 1935.

70. **Tutin, T. G.,** The antecology of *Zostera marina* in relation to its wasting disease, *New Phytol.,* 37, 50, 1938.

71. **Stevens, N. E.,** Environmental conditions and the wasting disease of eelgrass, *Science,* 84, 87, 1936.

72. **Stevens, N. E.,** Environmental factors and the wasting disease of eelgrass, *Rhodora,* 41, 260, 1939.

73. **Taylor, J. L. and Saloman, C. H.,** Some effects of hydraulic dredging and coastal development in Boca Ciega Bay, Florida, *Fish. Bull.,* 67, 213, 1968.

74. **Godcharles, M. F.,** A study of the effects of a commercial hydraulic clam dredge on benthic communities in the estuarine areas, *Fla. Dep. Nat. Resour. Div. Mar. Resour. Tech. Ser.,* 64, 1, 1971.

75. **Zieman, J. C.,** The ecological effects of physical damage from motor boats on turtlegrass beds in southern Florida, *Aquat. Bot.,* 2, 127, 1976.

76. **Camp, D. K., Cobb, S. P., Van Breedveld, J. F.,** Overgrazing of seagrasses by a regular urchin, *Lytechinus variegatus, Bioscience,* 23, 37, 1973.

77. **Orth, R. J.,** The importance of sediment stability in seagrass communities, in *Ecology of Marine Benthos,* Coull, B. C., Ed. University of South Carolina Press, Columbia, 1977, 122.

78. **Roessler, M. A. and Zieman, J. C.,** The effects of thermal additions on the biota of Southern Biscayne Bay, Florida, in *Proc. 22nd Ann. Session Gulf Caribb. Fish. Inst.,* 1969, 136.

79. **Blake, N. J., Doyle, L. J., and Pyle, T. E.,** An ecological study in the vicinity of the P. L. Bartow Power Plant, Tampa Bay, Florida, Final Rep., University of South Florida, Tampa, Fla., 1974.

80. **Zieman, J. C.,** Tropical seagrass ecosystems and pollution, in *Tropical Marine Pollution,* Vol. 12, Wood, E. J. F. and Johannes, R. C., Eds., Elsevier Oceanogr. Ser., Amsterdam, 1975, 63.

81. **McNulty, J. K.,** Ecological effects of sewage pollution in Biscayne Bay, Florida sediments and the distribution of benthic and fouling macroorganisms, *Bull. Mar. Sci.,* 11, 394, 1970.

82. **Taylor, J. L. and Prest, K. W.,** Harvest and regrowth of turtlegrass (*Thalassia testudininum*) in Tampa Bay, Florida, *Fish. Bull.,* 67, 123, 1968.

83. **Lewis, R. R., Phillips, R. C., and Lombardo, R.,** Seagrass mapping project, Hillsborough County, Florida, *Fla. Sci.,* 44 (Suppl.), 25, 1981.

84. **Dong, M., Rosenfeld, J., Redmann, G., Elliott, M., Balazy, J., Poole, B., Ronnhalm, K., Kenisberg, D., Novak, P., Cunningham, C., and Karnow, C.,** The role of man-induced stresses in the ecology of Long Reef and Christiansted Harbor, St. Croix, U.S. Virgin Islands, West Indies Laboratory, Fairleigh Dickinson University, St. Croix, Spec. Publ., 1972.

85. **Nichols, M. M., Van Eepoel, R., Grigg, D., Brady, R., Sallenger, A., Olman, J., and Crena, R.,** Environment, water and sediments of Christiansted Harbor, St. Croix, Division of Environmental Health, Government of the Virgin Islands, Rept. No. 16, 1972.

86. **Verquet, J. and Beaux, G.,** *Jardinier de la Mer,* L'Association — Fondation G. Cooper, Hyeres, France, 1976, cahier No. 1.

87. **Larkum, A. W. D.,** Ecology of Botany Bay, I. Growth of *Posidonia australis* (Brown) Hoak. f. in Botany Bay and other bays of the Sydney Basin, *Aust. J. Mar. Freshwater Res.,* 27, 117, 1976.

88. **Faraday, W. E. and Churchill, A. C.,** Uptake of cadmium by the eelgrass *Zostera marina, Mar. Biol.,* 53, 293, 1979.

89. Penello, W. F. and Brunkhuis, B. H., Cadmium and manganese flux in eelgrass *Zostera marina* I. Modeling dynamics of metal release from labelled tissues, *Mar. Biol.*, 58, 181, 1980.

90. Brinkhuis, B. H., Penello, W. F., and Churchill, A. C., Cadmium and manganese flux in eelgrass *Zostera marina* II. Metal uptake by leaf and root-rhizome tissues, *Mar. Biol.*, 58, 187, 1980.

91. Parker, P. L., Gibbs, A., and Lowler, R., Cobalt, iron, and manganese in a Texas bay, *Publ. Inst. Mar. Sci. Univ. Tex.*, 928, 1963.

92. Parker, P. L., Zinc in a Texas bay, *Publ. Inst. Mar. Sci. Univ. Tex.*, 8, 75, 1962.

93. Stevenson, R. A. and Ufret, S. L., Iron, manganese, and nickel in skeletons and food of the sea urchin, *Tripneustes esculentus* and Echinometra *lucunter*, *Limnol. Oceanogr.*, 11, 11, 1966.

94. Thomas, M. L. H., Experimental control of eelgrass (*Zostera marina* L.) in oyster growing areas, in *Proc. Northeast. Weed Control Conf.*, 21, 542, 1967.

95. Zobell, C. D., The occurrence, effects and fate of oil polluting the sea, *Int. J. Air Water Pollut.*, 7, 173, 1963.

96. Diaz-Pfirrer, M., The effects of an oil spill on the shore of Guanica, Puerto Rico, Assoc. Island Marine Labs, 4th Meet., Curaco, 1962.

97. Foster, M., Neushul, M., and Zingmark, R., The Santa Barbara oil spill, Part 2: initial effects on intertidal and kelp bed organisms, *Environ. Pollut.*, 2, 115, 1971.

98. Neushul, M., The effects of pollution on populations of intertidal and subtidal organisms in southern California, in Santa Barbara Oil Spill Symposium, University of California, Santa Barbara, December 16—18, 1970.

99. Dalby, D. H., Some factors controlling plant growth in the intertidal environment, in *The Biological Effects of Oil Pollution on Littoral Communities,* Carthy, J. B. and Arthur, D. R., Eds., Field Studies 2 (Suppl.), 1968, 21.

100. Jacobs, R. P. W. M., Effects of the 'Amoco Cadiz' oil spill on the seagrass community at Roscoff with special reference to the benthic infauna, *Mar. Ecol., Prog. Ser.*, 2, 207, 1980.

101. Moffitt, J. and Cottam, C., Eelgrass depletion on the Pacific coast and its effect upon black brant, U.S. Fish and Wildlife Service Leaflet, 1941, 204.

102. Brijns, M. F. M. and Tanis, J., De rotganzen, *Branta bernicla* (L.) opter schelling, *Ardea,* 43, 261, 1955.

103. Ranwell, D. S. and Downing, B. M., Brant goose winter feeding pattern and Zostera resources at Scott Head Island, Norfolk, *Anim. Behav.*, 7, 42, 1959.

104. Dexer, R. W., Ecological significance of the disappearance of eelgrass at Cape Ann, Massachusetts, *J. Wildl. Manage.*, 8, 1973, 1944.

105. Marshall, N., Abundance of bay scallops in the absence of eelgrass, *Ecology,* 28, 321, 1967.

106. Thayer, G. W. and Stewart, H. H., The bay scallop makes its bed of eelgrass, *Mar. Fish. Rev.,* 36, 27, 1974.

107. Stauffer, R. C., Changes in the invertebrate community of a lagoon after disappearance of the eelgrass, *Ecology,* 18, 427, 1937.

108. Wilson, D. P., The decline of *Zostera marina* L. at Salcombe and its effects on the shore, *J. Mar. Biol. Assoc. U.K.*, 28, 295, 1949.

109. Peterson, C. G. J., The sea bottom and its production of fish food. A survey of the work done in connection with valuation of the Danish waters from 1883—1917, *Rep. Dan. Biol. Stn.*, 25, 1, 1918.

110. Springer, V. G. and Woodburn, K. D., An ecological study of the fishes of Tampa Bay area, *Fla. State Board Conserv. Mar. Lab., Prof. Paper Ser.*, 1, 1, 1960.

111. Phillips, R. C., Observations on the ecology and distribution of the Florida seagrasses, *Fla. State Board Conserv. Mar. Lab., Prof. Paper Ser.*, 2, 1, 1960.

112. Phillips, R. C., Responses of transplanted and indigenous *Thalassia testudinum* Banks ex König and *Halodule wrightii* Aschers. to sediment loading and cold stress, *Contr. Mar. Sci.*, 23, 79, 1980.

113. Cooper, G., Souchon, L., Verguet, J., and Beaux, G., *Jardinier de la Mer,* L'Association — Fondation G. Cooper, Hyeres, France, 1977, cahier No. 2.

114. Zieman, J. C., The effects of a Thermal Effluent Stress on the Seagrasses and Macroalgae in the Vicinity of Turkey Point, Biscayne Bay, Florida, Ph.D. thesis, University of Miami, Coral Gables, Fla., 1970.

115. Roessler, M. A. and Zieman, J. C., The effects of thermal additions on the biota of Southern Biscayne Bay, Florida, in *Proc. 22nd Ann. Sess. Gulf Caribb. Fish. Inst.*, Miami, 1969, 136.

116. Thorhaug, A., The flowering and fruiting of restored *Thalassia* beds: a preliminary note, *Aquat. Bot.*, 6, 189, 1979.

117. Jacobs, R. P. W. M., Distribution and aspects of the production and biomass of eelgrass, *Zostera marina, Aquat. Bot.*, 7, 151, 1979.

118. Aioi, K., Mukai, H., Koike, I., Ontsu, M., and Hattori, A., Growth and organic production of eelgrass (*Zostera marina* L.) in temperate waters of the Pacific coast of Japan. II. Growth analysis in winter, *Aquat. Bot.*, 10, 175, 1981.

119. **Mukai, H., Aioi, K., Koike, I., Iizumi, H., Ohtsu, M., and Hattori, A.,** Growth and organic production of eelgrass (*Zostera marina* L.) in temperate waters of the Pacific coast of Japan. I. Growth analysis in spring — summer, *Aquat. Bot.,* 7, 47, 1979.

120. **Ott, J. A.,** Growth and production in *Posidonia oceanica* (L.) Delile, Mar. *Ecology,* 1, 47, 1980.

121. **Thomas, M. L. H.,** Control of eelgrass using 2, 4-D in oyster-growing areas, Fish. Res. Bd. Can. Mar. Ecol. Lab., Bedford Inst., Dartmouth, Nova Scotia, General Ser. Circ. No. 1, 1968.

122. **Short, F. T.,** Eelgrass Production in Charlestown Pond: An Ecological Analysis and Numerical Simulation Mode 1, M.S. thesis, Graduate School Oceanography, University of Rhode Island, Kingston, 1975.

123. **Littler, M. M. and Murray, S. N.,** The primary productivity of marine macrophytes from a rocky intertidal community, *Mar. Biol.,* 27, 131, 1974.

124. **Williams, S. L.,** Seagrass Productivity: The Effect of Light on Carbon Uptake, M.S. thesis, University of Alaska, Fairbanks, 1977.

125. **Pomeroy, L. R.,** Primary productivity of Boca Ciega Bay, Florida, *Bull. Mar. Sci. Gulf Caribb.,* 10, 1, 1960.

126. **Buesa, R. J.,** Population and biological data on turtle grass (*Thalassia testudinum* König) on the northwestern Cuban shelf, *Aquaculture,* 4, 199, 1974.

127. **Cooper, G.,** *Posidonia* et la desertification sous la mer, *Peuples Mediterr.,* 4, 43, 1978.

128. **Bittaker, H. F.,** A Comparative Study of the Phytoplankton and Benthic Macrophyte Primary Productivity in a Polluted Versus an Unpolluted Coastal Area, M.S. thesis, Florida State University, Tallahassee, 1975.

129. **Van Eepoel, R. O., Griff, D. I., Brady, R. W., and Raymond, W.,** Water quality and sediments in Lindbergh Bay, St. Thomas, Caribb. Res. Inst. Water Pollut. Rep. II, 1971.

130. **Grigg, D. I., Shatrosky, E. L., and Van Eepoel, R. P.,** Operating efficiencies of package sewage plants on St. Thomas, Virgin Islands, August to December 1970, Caribb. Res. Inst. Water Pollut. Rep. No. 12, 1971.

131. **Zimmerman, M. S. and Livingston, R. J.,** Effects of Kraft mill effluents on benthic macrophyte assemblages in a shallow bay system, Apalachee Bay, North Florida, U.S.A., *Mar. Biol.,* 34, 297, 1976.

132. **Phillips, R. C.,** Transplanting methods, in *Handbook of Seagrass Biology: An Ecosystem Perspective,* Phillips, R. C. and McRoy, C. P., Eds., Garland STPM Press, New York, 1980, chap. 4.

133. **Ranwell, D. S., Wyer, D. W., Boorman, L. A., Pizzey, J. M., and Waters, R. J.,** *Zostera* transplants in Norfolk and Suffolk, Great Britain, *Aquaculture,* 4, 185, 1974.

134. **Churchill, C. A., Cok, A. E., and Riner, M. I.,** Stabilization of subtidal sediments by the transplantation of the seagrass *Zostera marina* L., New York Sea Grant, 1978, Rept. No. NYSSGP-RS-78-15.

135. **Phillips, R. C.,** Creation of seagrass beds, in Rehabilitation and Creation of Selected Coastal Habitats: Proc. workshop, Lewis, J. C. and Bunce, E. W., Eds., Office of Biological Services, U.S. Fish and Wildlife Service, Washington, D.C., 1980.

136. **Phillips, R. C.,** Planting guidelines for seagrasses, Coastal Engineering Tech. Aid No. 80-2, U.S. Army, Corps of Engineers, Coastal Eng. Res. Cent., Fort Belvoir, Va., 1980.

137. **Thorhaug, A. and Austin, C. B.,** Restoration of seagrasses with economic analysis, *Environ. Conserv.,* 3, 259, 1976.

138. **van Breedveld, J. F.,** Transplanting of seagrasses with emphasis on the importance of substrate, *Fla. Mar. Res. Publ.,* 17, 1, 1975.

139. **Kelly, J. A., Fuss, C. M., and Hall, J. R.,** The transplanting and survival of turtle grass, *Thalassia testudinum,* in Boca Ciega Bay, Florida, *Fish. Bull.,* 69, 273, 1971.

140. **Thorhaug, A. and Hixon, R.,** Revegetation of *Thalassia testudinum* in a multiple-stressed estuary, North Biscayne Bay, Florida, in *Proc. 2nd Annu. Conf. Restor. Coastal Vegetation Fla.,* Hillsborough Community College, Tampa, Fla., 1975, 12.

141. **Carangelo, P. D., Oppenheimer, C. H., and Picarazzi, P. E.,** Biological application for the stabilization of dredged materials, Corpus Christi, Texas: submergent plantings, in *Proc. 6th Annu. Conf. Restor. Coastal Vegetation Fla.,* Hillsborough Community College, Tampa, Fla., 1979.

142. **Eleuterius, L. N.,** A study of plant establishment on spoil areas in Mississippi Sound and adjacent waters, Final Rep. to Corps of Engineers, U.S. Army, Gulf Coast Res. Lab., Ocean Springs, Miss., 1974.

143. **Phillips, R. C.,** Transplantation of seagrasses, with special emphasis on eelgrass, *Zostera marina* L., *Aquaculture,* 4, 161, 1974.

144. **Rogers, R. G. and Bisterfeld, F. T.,** Seagrass vegetation attempts in Escambia Bay, Florida, during 1974, in *Proc. 2nd Ann. Conf. Restor. Coastal Vegetation Fla.,* Hillsborough Community College, Tampa, Fla., 1975.

145. **Phillips, R. C., Vincent, M. K., and Huffman, R. T.,** Habitat development field investigations, Port St. Joe — Seagrass Demonstration site, Port St. Joe, Florida, Tech. Rep. D-78-33, U.S. Army Engineer Waterways Rep. D-78-33, U.S. Army Eng. Waterways Exp. Stn., Vicksburg, Miss., 1978.

146. **Addy, C. D. and Aylward, D. A.**, Status of eelgrass in Massachusetts during 1943, *J. Wildl. Manage.*, 8, 269, 1944.

147. **Cottam, C. and Addy, C. E.**, Present eelgrass condition and problems on the Atlantic coast of North America, 12th North Am. Wildl. Conf., San Antonio, Texas, 1947.

148. **Cottam, C. and Munro, D. A.**, Eelgrass status and environmental relations, *J. Wildl. Manage.*, 18, 449, 1944.

149. **Fonseca, M. S., Kenworthy, W. J., Homziak, J., and Thayer, G. W.**, Transplanting of eelgrass and shoalgrass as potential means of economically mitigating a recent loss of habitat, in *Proc. 6th Am. Conf. Restor. Coastal Vegetation Fla.*, Hillsborough Community College, Tampa, Fla., 1979.

150. **Thorhaug, A.**, Transplantation techniques for the seagrass *Thalassia testudinum,* Tech. Bull., No. 34, University of Miami Sea Grant, Coral Gables, Fla., 1976.

151. **Kenworthy, W. J. and Fonseca, M. S.**, Reciprocal transplant of the seagrass *Zostera marina* L. Effect of substrate on growth, *Aquaculture,* 12, 197, 1977.

152. **Thorhaug, A.**, Transplantation of the seagrass *Thalassia testudinum* König, *Aquaculture,* 4, 177, 1974.

153. **Cooper, G., Verquet, J., and Souchon, L.**, *Jardinier de lar Mer,* L'Association — Fondation G. Cooper, Hyeres, France, 1979, cahier No. 3.

154. **Cooper, G.**, *Jardinier de la Mer,* L'Association — Foundation G. Cooper, Hyeres, France, 1981, cahier No. 5.

155. **Cooper, G.**, *Jardinier de la Mer,* L'Association — Fondation G. Cooper, Hyeres, France, 1981, cahier No. 5.

156. **Robilliard, G. A. and Porter, P. E.**, Transplantation of eelgrass (*Zostera marina*) in San Diego Bay, Undersea Sciences Department, Naval Undersea Center, San Diego, California, 1976.

157. **Addy, C. E.**, Eelgrass Planting guide, *Md. Conserv.,* 24, 16, 1947.

158. **Addy, C. E.**, Germination of eelgrass seed, *J. Wildl. Manage.,* 11, 279, 1947.

159. **Hatade, K., Ueki, Y., and Kunitake, K.**, Studies on propagation of eelgrass, *Zostera marina* — II. Mass collection of seeds and sowing in the field, *Stud. Tech. Aquaculture,* 4, 7, 1975.

160. **Hatade, K., Ogawa, K., and Kunitake, K.**, Studies on propagation of eelgrass, *Zostera marina* — III. Building of eelgrass bed by seed - sowing, *Stud. Tech. Aquaculture,* 4, 21, 1975.

161. **Hatade, K., Ogawa, K., and Kunitake, K.**, Studies on propagation of eelgrass, *Zostera marina* — IV. Building of eelgrass bed by seed-sowing and rhizome transplanting, *Stud. Tech. Aquaculture,* 5, 17, 1976.

162. **Harris, M. M., King, R. J., and Ellis, J.**, The eelgrass *Zostera capricorni* in Illawarra Lake, New South Wales, *Proc. Linn. Soc. N.S.W.,* 104, 23, 1980.

163. **Harris, M. M.**, Ecological Studies on Illawarra Lake with Special Reference to *Zostera capricorni* Aschers., M.S. thesis, University of New South Wales, Australia, 1977, 183.

164. **Penn, N.**, personal communication, 1980.

165. **Peres, J. M.**, personal communication, 1974.

166. **Lewis, R. R.**, personal communication, 1980.

167. **Backman, T. W.**, unpublished data, 1973.

168. **Phillips, R. C.**, unpublished data, 1980.

169. **Jones, R. D.**, unpublished data, 1965.

170. **Lewis, R. R.**, unpublished data, 1981.

171. **Bittaker, H. F.**, unpublished data, 1980.

Appendix I

# COASTAL PLANT MATERIAL SUPPLIERS

## U.S.

### Marsh Plants

Dune and Marsh, Inc.
1511 Dixie Trail
Raleigh, N.C. 27607
(919) 787-5110, 458-8498, 737-3288

San Francisco Bay Marine Research
  Center
8 Middle Road
Lafayette, Calif. 94549
(415) 332-5100

Environmental Concern, Inc.
P.O. Box P
St. Michaels, Md. 21663
(301) 745-9620

Horticultural Systems, Inc.
P.O. Box 70
Parrish, Fla. 33564
(813) 776-1605

Mangrove Systems, Inc.
P.O. Box 24748
Tampa, Fla. 33623
(813) 257-3231

Environmental Wetland Gardens
10913 Sheldon Road
Tampa, Fla. 33624
(813) 920-6259

Ecoshores, Inc.
585 Beville Road
Daytona Beach, Fla. 32019
(904) 767-6200

Paul D. Carangelo
P.O. Box 1361
Port Aransas, Tex. 78373
(512) 749-5477, 937-4873

### Dune Plants

Allen DeVries
14835 Barry
Holland, Mich. 49423

Church's Greenhouse and Nursery
Old Shore Road
Erman Road, # 1
Cape May, N.J. 08204

Jackson Seed Company
East 8803 Sprague Ave.
Spokane, Wash. 99206

Mason-Lake Soil Conservation Districts
102 East 5th St.
Scottsville, Mich. 49454

Muskegon Soil Conservation District
Federal Building, Room 207
Muskegon, Mich. 49443

Van Pine, Inc.
Route # 1
West Olive, Mich. 49460

Wilbur Ternyik
921 Rhododendron
Florence, Ore. 97439

Seacoast Plant Nursery
c/o Karl Graetz
Rt. 1, Box 874
Morehead, N.C. 28557

Mangrove Systems, Inc.
P.O. Box 24748
Tampa, Fla. 33623
(813) 257-3231

Bar-Don Nursery
2611 Lockmoe Drive
Raleigh, N.C. 27608

Coastal Stabilization Nursery
Box 65
New Bern, N.C. 25860

Manistee Soil Conservation District
P.O. Box 275
Onekama, Mich. 49675

Moores Sod Farm
P.O. Box 376
Berlin, Md. 21811

R & R Beachgrass
Box 33-R, D. I.
Lewes, Del. 19958

D. M. Bryan
Route 2
Garner, N.C. 27529

Dune and Marsh, Inc.
1511 Dixie Trail
Raleigh, N.C. 27607
(919) 787-5110, 458-8498, 737-3288

Horticultural Systems, Inc.
P.O. Box 70
Parrish, Fla. 33564

Seacoast Enterprises
Rt. 1, Box 129
Newport, N.C. 28570

## Mangroves

Horticultural Systems, Inc.
P.O. Box 70
Parrish, Fla. 33564
(813) 776-1605

Mangrove Systems, Inc.
P.O. Box 24748
Tampa, Fla. 33623
(813) 257-3231

Howard J. Teas Nursery
6700 S.W. 130th Terrace
Miami, Fla. 33156
(305) 284-4125

Tropical Bioindustries
Development Company
9000 S.W. 87th Court
Suite 104
Miami, Fla. 33156
(305) 279-7026

Pine Breeze Nursery
P.O. Box 3
Bokeelia, Fla. 33922
(813) 283-2385 or 955-3553

Herren Nursery
Rt. 2, Box 142
Lake Placid, Fla. 33852
(813) 465-0024

SCCF Native Plant Nursery
P.O. Drawer S
Sanibel, Fla. 33957
(813) 472-1932

Florida Keys Native Nursery, Inc.
102 Mohawk Street
Tavernier, Fla. 33070
(305) 852-5515

Associated Marine Institutes, Inc.
6850 Benjamin Road
Tampa, Fla. 33614
(813) 885-6918

Tropical Greenery & Native Nursery
22140 SW 152nd Avenue
Goulds, Fla. 33170
(305) 248-5529

Environmental Wetland Gardens
10913 Sheldon Road
Tampa, Fla. 33624
(813) 920-6259

Ecoshores, Inc.
585 Beville Road
Daytona Beach, Fla. 32019
(904) 767-6200

Dr. Peter Schroeder
11550 S.W. 108 Ct.
Miami, Fla. 33176
(305) 238-5509

Dr. Calvin McMillan
Department of Botany
University of Texas at Austin
Austin, Tex. 78712

## Seagrasses

Mangrove Systems, Inc.
P.O. Box 24748
Tampa, Fla. 33623
(813) 257-3231

Applied Marine Ecological Service, Inc.
600 Grapetreee Drive, Suite 4EN
Key Biscayne, Fla. 33149
(305) 361-3340

## PUERTO RICO

Pennock Gardens
G.P.O. Box 3587
San Juan, PR 00936
(809) 785-8080

Mangrove Systems, Inc.
P.O. Box 24748
Tampa, Fla. 33623
(813) 257-3231

Howard J. Teas Nursery
6700 S.W. 130th Terrace
Miami, Fla. 33156
(305) 284-4125

## GREAT BRITAIN/EUROPE

Dr. Derek Ranwell
School of Biological Sciences
University of East Anglia
Norwich NR4 7TJ
England

Dr. Ernest Seneca
Botany Department
North Carolina State University
Raleigh, N.C., 27650

## AUSTRALIA

Contact:  Coastal Wetlands Society
P.O. Box A225
Sydney South 2000
Australia
for possible suppliers

Appendix II

## PROFESSIONAL SOCIETIES

Society of Wetlands Scientists
P.O. Box 296
Wilmington, N.C. 28402
*U.S.*
$20/year

Natural Areas Association: 320 South
 Third St.
Rockford, Ill. 61108
*U.S.*
$10/year

Coast and Wetlands Society
P.O Box A225
Sydney South 2000
New South Wales
*Australia*
$7.50/year

## PROFESSIONAL JOURNALS

Restoration and Management Notes
University of Wisconsin-Madison
 Arboretum
1207 Seminole Highway
Madison, Wis. 53711
*U.S.*
$8/year

National Wetlands Newsletter
Environmental Law Institute
Suite 600
1346 Connecticut Ave., N.W.
Washington, D.C. 20036
*U.S.*
$21/year

## PROFESSIONAL MEETINGS

Wetlands Creation and Restoration
 Conference
Environmental Studies Center
Hillsborough Community College
P.O. Box 30030
Tampa, Fla. 33630
*U.S.*
Held annually since 1974 — May of
 each year. Complete listing of previous
 proceedings issues and availability can
 be requested from:
   Roy R. "Robin" Lewis, III
   Wetlands Conference Convenor
   P.O. Box 24748
   Tampa, Fla. 33623

Mitigation Symposium
The Rocky Mountain Forest and Range
 Experiment Station
240 W. Prospect St.
Fort Collins, Colo. 80526
*U.S.*
First symposium held in July 1979. Free
 copies of proceedings are available
 from the above address.

# INDEX